SIMPLIFIED DESIGN OF CONCRETE STRUCTURES

Sixth Edition

THE LATE HARRY PARKER, M.S.
*Formerly Professor of Architectural Construction
University of Pennsylvania*

prepared by

JAMES AMBROSE, M.S.
Professor of Architecture
University of Southern California

A WILEY-INTERSCIENCE PUBLICATION

JOHN WILEY & SONS, INC.

New York • Chichester • Brisbane • Toronto • Singapore

Copyright © 1991 by John Wiley & Sons, Inc.

Library of Congress Cataloging-in-Publication Data:

Parker, Harry, 1887–
　　Simplified design of concrete structure / Harry Parker. — 6th ed.
　/ prepared by James Ambrose.
　　　　p.　　cm. — (Parker-Ambrose series of simplified design guides)
　　"A Wiley-Interscience publication."
　　Includes bibliographical references.
　　　1. Reinforced concrete construction.　I. Ambrose, James E.
II. Title.　III. Series.
TA683.2.P3　　1991
693'.54—dc20　　　　　　　　　　　　　　　　　　　　　90-37690
ISBN 0-471-52204-X　　　　　　　　　　　　　　　　　　　CIP

Printed in the United States of America

10 9 8 7 6 5 4 3 2 1

CONTENTS

PREFACE TO THE SIXTH EDITION

This new edition of Professor Parker's enduringly popular book has been prepared to bring the material into conformance with current codes, design standards, and industry practices. New materials have been added to develop a form relating to the recent new editions of other books in Professor Parker's series on simplified design. A major addition consists of the new Chapter 12, containing design examples of structural systems for three different buildings. Other new materials in this edition include treatments of concrete frames, tilt-up walls, and structural masonry with concrete units.

Materials contained in the previous edition have generally been retained, with any necessary revisions. The topic coverage and style of the work remain the same, and the new materials are meant to improve the usefulness of the book, not to change its general character. The principal purposes of the book are as stated in Professor Parker's Preface to the First Edition, which follows.

Readers using the book in a self-study situation will find the Study Aids at the back of the book to be useful in gauging general comprehension of the materials. Use of these, plus the exercise problems, should serve to engage the reader in effective study of the subject. Answers to both general questions and numerical exercises are given for self-evaluation.

I am grateful to the International Conference of Building Officials, publishers of the Uniform Building Code, and to the American Concrete Institute for permission to use materials from their publications. As my work is done almost entirely in my home, I am also grateful to my family for their indulgence, their patience, and their frequent assistance.

JAMES AMBROSE

Westlake Village, California
October 1990

PREFACE TO THE
FIRST EDITION

The preparation of this book has been prompted by the fact that many young men desirous of the ability to design elementary reinforced concrete structural members have been deprived of the usual preliminary training. The author has endeavored to simplify the subject matter for those having a minimum of preparation. Throughout the text will be found references to Section I of *Simplified Engineering for Architects and Builders*. Familiarity with this brief treatment of the principles of mechanics is sufficient. Any textbook on mechanics will give the desired information. This particular book has been referred to as a convenience in having a direct reference. With these basic principles, and a high school knowledge of algebra, no other preparation is needed.

In preparing material for this book, the author has had in mind its use as a textbook as well as a book to be used for home study. Simple, concise explanations of the design of the most common structural elements have been given rather than discussions of the more involved problems. In addition to the usual design formulas sufficient theory underlying the principles of design is presented, in developing basic formulas, to ensure the student a thorough knowledge of the fundamentals involved.

A major portion of the book contains illustrative examples giving the solution of the design of structural members. Accompanying the examples are problems to be solved by the student. The usual tables necessary in the design of reinforced concrete are included. No supplementary books, tables, or charts are required. Where practicable, safe load tables have been added, but in each instance illustrative examples give the design steps showing the underlying principles by means of which the table was prepared.

The titles of other volumes in this series of elementary books relating to structural design have included the words "for Architects and Builders." The purpose of this has been to convey the idea that the books are limited in scope and that they are not the comprehensive and thorough treatises demanded by engineers. It is found, however, that these books have a much wider use than was anticipated. Because of this, it has seemed advisable to omit the words "for Architects and Builders" from the title of the present volume even though it has been prepared with this particular group in mind.

Grateful appreciation is extended to the Portland Cement Association and the American Concrete Institute for their kindness and cooperation in granting permission to reproduce data and tables from their publications.

The author has made no attempt to offer short-cuts or originality in design. Instead, he has endeavored to present clearly and concisely the present day methods commonly used in the design of reinforced concrete members. A thorough knowledge of the principles herein set forth should encourage the student and serve as adequate preparation for advanced study.

HARRY PARKER

High Hollow, Southampton, Pa.
March 1943

SIMPLIFIED DESIGN OF
CONCRETE STRUCTURES

1

INTRODUCTION

Concrete is a widely used, diverse material. This chapter presents some of the considerations for the use of concrete for structural purposes in building construction.

1.1 CONCRETE AS A STRUCTURAL MATERIAL

Concrete consists of a mixture that contains a mass of loose, inert particles of graded size (commonly sand and gravel) held together in solid form by a binding agent. That general description covers a wide range of end products. The loose particles may consist of wood chips, industrial wastes, mineral fibers, and various synthetic materials. The binding agent may be coal tar, gypsum, portland cement, or various synthetic compounds. The end products range from asphalt pavement, insulating fill, shingles, wall panels, and masonry units to the familiar sidewalks, roadways, foundations, and building frameworks.

This book deals primarily with concrete formed with the common binding agent of *portland cement*, and a loose mass consist-

1

ing of sand and gravel. This is what most of us mean when we use the term concrete. With minor variations, this is the material used mostly for structural concrete—to produce building structures, pavements, and foundations.

Concrete made from natural materials was used by ancient builders thousands of years ago. Modern concrete, made with industrially produced cement, was first produced in the early part of the nineteenth century when the process for producing portland cement was developed. Because of its lack of tensile strength, however, concrete was used principally for crude, massive structures—foundations, bridge piers, and heavy walls.

In the late nineteenth century, several builders experimented with the technique of inserting iron or steel rods into relatively thin structures of concrete to enhance their ability to resist tensile forces. This was the beginning of what we now know as *reinforced concrete*. Many of the basic forms of construction developed by these early experimenters have endured to become part of our common technical inventory for building structures.

Over the years, from ancient times until now, there has been a steady accumulation of experience derived from experiments, research, and, mostly recently, intense development of commercial products. As a result, there is presently available to the building designer an immense variety of products under the general classification of concrete. This range is somewhat smaller if major structural usage is required, but the potential variety is still significant.

1.2 FORMS OF CONCRETE STRUCTURES

For building structures, concrete is mostly used with one of three basic construction methods. The first is called *sitecast concrete*, in which the wet concrete mix is deposited in some forming at the location where it is to be used. This method is also described as *cast-in-place* or *in situ* construction.

A second method consists of casting portions of the structure at a location away from the desired location of the construction. These elements—described as *precast concrete*—are then moved into position, much as blocks of stone or parts of steel frames are.

Finally, concrete may be used for masonry construction—in one of two ways. Precast units of concrete [called concrete masonry units (CMUs)] may be used in a manner similar to bricks or stones. Or, concrete fill may be used to produce solid masonry by being poured into cavities in masonry produced with bricks, stone, or CMUs. The latter technique, combined with the insertion of steel reinforcement into the cavities, is widely used for masonry structures today. The use of concrete-filled masonry, however, is one of the oldest forms of concrete construction—used extensively by the Romans and the builders of early Christian churches.

Concrete is produced in great volume for various forms of construction. Building frames, walls, and other structural systems represent a minor usage of the total concrete produced. Pavements for sidewalks, parking lots, streets, and ground-level floor slabs in buildings use more concrete than all the building frameworks. Add the usage for the interstate highway system, water control, marine structures, and large bridges and tunnels, and building structural usage shrinks into insignificance. One needs to understand this when dealing with considerations of the economics and operation of the concrete industry.

Other than for pavements, the widest general use of concrete for building construction is for foundations. Almost every building has a concrete foundation, whether the major aboveground construction is concrete, masonry, wood, steel, aluminum, or fabric. For small buildings with shallow footings and no basement, the total foundation system may be modest, but for large buildings and those with many below-ground levels, there may well be a gigantic underground concrete structure.

For above-ground building construction, concrete is generally used in situations that fully realize the various advantages of the basic material and the common systems that derive from it. For structural applications, this means using the major compressive resistance of the material and in some situations its relatively high stiffness and inertial resistance (major dead weight). However, in many applications, the nonrotting, vermin- and insect-resistive, and five-resistive properties may be of major significance. And for many uses, its relatively low bulk volume cost is important.

1.3   WORK IN THIS BOOK

As the lead word of the title implies, the work in this book is of a limited form with regard to the potential complexity of the topic. The limitations have to do mostly with the level of complexity of the problems, the structural components and systems, and the type of mathematical computations presented. The basic intent is to keep the presentations within the potential grasp of persons with limited backgrounds in mathematics, applied mechanics, fundamentals of engineering investigation, and general, problem-solving, computational skills.

Computations for the investigation of structural behaviors are a critical necessity for serious structural design work, but in fact constitute only a small fraction of the total work of structural design. Structural planning, integration of the structure with other building subsystems, development of construction details, and the overall decision-making processes of design development draw on knowledge and skills beyond those of mathematics and applied physics.

The work presented here seeks to span the range of concerns that impinge on the design of concrete structures for buildings. To appreciate the need for, and purposes of, structural computation, it is useful to perform some actual computational work. To that end, the reader is given many opportunities for involvement in the form of example computations in the text, with accompanying exercise problems of a similar form.

It is assumed that the reader will spend considerable time in a self-study mode, whether working in a structured course with an instructor or not. General presentations in the text, opportunities for attempting exercise problems, and the supplemental study aids at the end of the book are intended to support this mode of study. Answers to all the exercise problems and questions are provided for self-checks.

While all of the data sources and general reference materials for the work in this book are provided here, more benefit will be gained if the reader has access to some additional references. The principal such item is the ACI Code (Ref. 1, at the end of the book). Several excerpts from that reference are included here, but general familiarity with the entire code is to be encouraged. For readers with a lack of such experience, some reference to a text on general building construction will also be useful for a broader treatment of concrete as a general building construction material.

For readers who intend to pursue the study of structural design of concrete beyond the scope of this book, it would be useful to have access to a more advanced text, which may provide some supplementary development of selected topics. Several good books exist for this basic purpose, one of which is listed in the Reference section of the book (Ref. 5).

1.4   REFERENCE SOURCES

Information about concrete structures is forthcoming from a number of sources. These range from relatively unbiased textbooks and research reports to clearly biased promotional materials from the producers and suppliers of construction materials, equipment, and services. Bias is not to be construed as evil here, it is merely to be acknowledged in evaluating the potential for completeness and neutrality in the materials presented. One cannot really expect people in the business of selling cement to provide unbiased information about the use of concrete, especially information about its drawbacks.

The total published information about concrete and all of the issues relating to it would fill a large building. Some is somewhat more essential and general in nature; most is highly detailed and narrowly directed. The material presented in this book is general in nature, specific to the interests of building designers, and a small, distilled essence of a number of general publications. Principal references used for this book are listed in the Reference section following the text.

There are several notable industry-wide organizations in the United States that have some relation to concrete. While these organizations are largely industry-supported, they do represent major sources of design codes and standards, as well as product information. They also perform or sponsor much of the research on which the design procedures are based. Materials from several of these sources are presented in this book, with the publications from which they are taken noted. Some of the major industry organizations are the following:

American Concrete Institute (ACI). This is an organization with a wide-ranging membership that includes major companies as well as many outstanding engineers, teachers, and researchers. One of its major publications is the ACI Code (full title: *Building Code Requirements for Reinforced Concrete*, ACI 318), which is widely used as the standard for design and construction of concrete structures.

Portland Cement Association (PCA). This organization is primarily sponsored by producers of cement, but conducts major research—much of it in its large testing facilities. It is a major producer of publications, including many significant research reports.

Concrete Reinforcing Steel Institute (CRSI). While steel reinforcement occupies a small volume in reinforced concrete, its cost is often close to or greater than that of the concrete, making it a major concern. The CRSI also sponsors research and produces many publications, a notable one being the *CRSI Handbook*.

Prestressed Concrete Institute (PCI). This organization publishes materials relating to industrially produced concrete structural elements and systems, as well as the general uses of prestressing and precasting.

National Concrete Masonry Association (NCMA). Masonry design and construction tend to be regionally specialized; thus there are many masonry industry organizations. They are generally friendly and cooperative with each other, but have some degree of regional limits. NCMA is a major national organization, although its headquarters are near Washington, D.C.

Masonry Institute of America (MIA). This organization is headquartered in Los Angeles, but has considerable influence on design and construction of reinforced masonry—a type of construction widely used in southern and western United States.

Anyone intending to pursue the study of this subject beyond the scope of the material in this book should consider obtaining a basic text, such as those used for courses in civil engineering schools. Some of these books currently in publication are listed in the Reference section at the end of the book.

1.5 DESIGN METHODS

There are two fundamentally different methods used for structural design. The traditional method, called the *working stress method*, is based on investigation of conditions that occur under the estimated actual usage conditions (called the *service loads*). Stresses and strains developed under these loads are investigated and compared to some level of permitted allowable stress or deformation. Safety in general is considered as the margin between the service load stresses and the ultimate strength of the materials. The allowable stresses are some fraction of the materials' limiting strength—the value of the fraction constituting the margin of safety.

For the working stress method, stresses and strains are in general investigated by using classical methods of structural analysis, mostly based on elastic behaviors. A major problem with the simple application of this method is that very few materials have pure elastic behavior for the entire range from no load to failure load. Thus various empirical adjustments must be made to derive meaningful allowable stresses. Safety becomes more abstract than the simple stress investigations imply.

The other basic method for structural design is variously called the *strength method*, the *ultimate strength method*, the *load factor method*, or the *factored load method*. Safety in this case is more directly considered by simple comparison of the service load with the estimated failure load of the structure. In general, the only stress conditions considered are those that occur at the failure of the structure. The strongest argument for use of this method is that the actual failure of a structure is relatively easy to test, and thus safety is more directly and positively assured. Use of the method with some assurance requires a reliance on the ability to predict failures of proposed designs, as it is not generally practical to test samples of all designs to failure. However, extensive testing of common elements has been done over the years, so that the necessary estimations of failure are quite reliable for ordinary construction.

Strength design was first developed mostly for use with reinforced concrete structures, and has been the general basis for requirements in the ACI Code for some time. The 1989 edition of the code, which is the edition used as a reference for this book, mentions the working stress method only as an "Alternative Design Method," presented in a small appendix section of the code. Strength design methods are now generally used in engineering design practice, except for some very minor structural elements. In fact, strength methods are now being generally applied to steel, wood, and masonry structures as well.

For study purposes and for very simple design tasks, the strength method constitutes a somewhat complex, empirical, and mystifying body of work. Individuals initially trained in classical elastic stress analysis usually approach this method by extension from that base. For purposes of explanation in this book, most situations will be presented in both methods, beginning in most cases with the simpler working stress analyses.

The working stress method is not recommended for design work in general, although it may be used in some cases for preliminary design estimates where simplified procedures may suffice. It may also be used with reasonable safety within the limits described in the ACI Code, but only for structures that comply with the following:

Minor construction in general (not multistory frames, long-span structures, etc.)

Low concrete strength (design strength not exceeding 4000 psi)

Low percentages of steel reinforcement (generally minimal reinforcement, not heavily reinforced structures).

Low-yield-strength reinforcement (yield strength not exceeding 60,000 psi)

1.6 STRUCTURAL COMPUTATIONS

The computational work presented in this book is mostly simple and can be performed easily with a pocket calculator. Structural computations for the most part can be rounded off. Accuracy beyond the third place is seldom significant, and many results presented in the work here have been so rounded off. In lengthy computations, however, it is advisable to carry one or more places of accuracy beyond that desired for the final answer. In the most part, the work shown here was performed on an eight-digit pocket calculator.

1.7 USE OF COMPUTERS

In most professional design firms structural computations done for final design work are performed with computer-aided procedures. Many standard programs are available for routine work and much of the necessary data is accessible from computer-retrievable sources. Many industry and professional organizations have software that can be purchased for ordinary design work.

Use of computer-aided methods permits faster accomplishment of tedious and complex investigations, more feasible study of alternatives, and design work that is interactive with that of others working on the same project. Many present design standards and codes have requirements and procedures that imply the use of computer-aided methods for practical design utilization.

The value of computer-aided methods increases with the level of complexity or sheer length of the computational work. For the most part, the work in this book is hardly worth doing with a computer. However, the purpose of this book is basically instructional, and hand operation of the full computational process allows for more involvement in the problems and the operation of their solutions.

1.8 UNITS OF MEASUREMENT

At the time of preparation of this edition, the building industry in the United States is still in a state of confused transition from the

TABLE 1.1   Units of Measurement: U.S. System

Name of Unit	Abbreviation	Use
Length		
Foot	ft	Large dimensions, building plans, beam spans
Inch	in.	Small dimensions, size of member cross sections
Area		
Square feet	ft^2	Large areas
Square inches	$in.^2$	Small areas, properties of cross sections
Volume		
Cubic feet	ft^3	Large volumes, quantities of materials
Cubic inches	$in.^3$	Small volumes
Force, Mass		
Pound	lb	Specific weight, force, load
KIP	kip	1000 pounds
Pounds per foot	lb/ft	Linear load (as on a beam)
Kips per foot	kips/ft	Linear load (as on a beam)
Pounds per square foot	lb/ft^2, psf	Distributed load on a surface
Kips per square foot	k/ft^2, ksf	Distributed load on a surface
Pounds per cubic foot	lb/ft^3, pcf	Relative density, weight
Moment		
Foot-pounds	ft-lb	Rotational or bending moment
Inch-pounds	in.-lb	Rotational or bending moment
Kip-feet	kip-ft	Rotational or bending moment
Kip-inches	kip-in.	Rotational or bending moment
Stress		
Pounds per square foot	lb/ft^2, psf	Soil pressure
Pounds per square inch	$lb/in.^2$, psi	Stresses in structures
Kips per square foot	$kips/ft^2$, ksf	Soil pressure
Kips per square inch	$kips/in.^2$, ksi	Stresses in structures
Temperature		
Degree Fahrenheit	°F	Temperature

TABLE 1.2 Units of Measurement: SI System

Name of Unit	Abbreviation	Use
Length		
Meter	m	Large dimensions, building plans, beam spans
Millimeter	mm	Small dimensions, size of member cross sections
Area		
Square meters	m^2	Large areas
Square millimeters	mm^2	Small areas, properties of cross sections
Volume		
Cubic meters	m^3	Large volumes
Cubic millimeters	mm^3	Small volumes
Mass		
Kilogram	kg	Mass of materials (equivalent to weight in U.S. system)
Kilograms per cubic meter	kg/m^3	Density
Force (Load on Structures)		
Newton	N	Force or load
Kilonewton	kN	1000 newtons
Stress		
Pascal	Pa	Stress or pressure (1 pascal = $1 \ N/m^2$)
Kilopascal	kPa	1000 pascal
Megapascal	MPa	1,000,000 pascal
Gigapascal	GPa	1,000,000,000 pascal
Temperature		
Degree Celsius	°C	Temperature

TABLE 1.3 Factors for Conversion of Units

To Convert from U.S. Units to SI Units, Multiply by:	U.S. Unit	SI Unit	To Convert from SI Units to U.S. Units, Multiply by:
25.4	in.	mm	0.03937
0.3048	ft	m	3.281
645.2	in.^{2}	mm^2	1.550×10^{-3}
16.39×10^3	in.^{3}	mm^3	61.02×10^{-6}
416.2×10^3	in.^{4}	mm^4	2.403×10^{-6}
0.09290	ft^2	m^2	10.76
0.02832	ft^3	m^3	35.31
0.4536	lb (mass)	kg	2.205
4.448	lb (force)	N	0.2248
4.448	kip (force)	kN	0.2248
1.356	ft-lb (moment)	N-m	0.7376
1.356	kip-ft (moment)	kN-m	0.7376
1.488	lb/ft (mass)	kg/m	0.6720
14.59	lb/ft (load)	N/m	0.06853
14.59	kips/ft (load)	kN/m	0.06853
6.895	psi (stress)	kPa	0.1450
6.895	ksi (stress)	MPa	0.1450
0.04788	psf (load or pressure)	kPa	20.93
47.88	ksf (load or pressure)	kPa	0.02093
$0.566 \times (°F - 32)$	°F	°C	$(1.8 \times °C) + 32$

use of English units (feet, pounds, etc.) to the new metric-based system referred to as the SI units (for Système International). Although a complete phase-over to SI units seems inevitable, at the time of this writing, construction-materials and products suppliers in the United States are still resisting it. Consequently, the ACI Code and most building codes and other widely used references are still in the old units. (The old system is now more appropriately called the U.S. system because England no longer uses it!) Although it results in some degree of clumsiness in the work, we have chosen to give the data and computations in this book in both units as much as is practicable. The technique is generally to perform the work in U.S. units and immediately follow it with the equivalent work in SI units enclosed in brackets [thus] for separation and identity.

Table 1.1 lists the standard units of measurement in the U.S. system with the abbreviations used in this work and a description of the type of the use in structural work. In similar form, Table 1.2 gives the corresponding units in the SI system. The conversion units used in shifting from one system to the other are given in Table 1.3.

For some of the work in this book, the units of measurement are not significant. What is required in such cases is simply to find a numerical answer. The visualization of the problem, the manipulation of the mathematical processes for the solution, and the quantification of the answer are not related to the specific units— only to their relative values. In such situations we have occasionally chosen not to present the work in dual units, to provide a less confusing illustration for the reader. Although this procedure may be allowed for the learning exercises in this book, the structural designer is generally advised to develop the habit of always indicating the units for any numerical answers in structural computations.

1.9 SYMBOLS

The following "shorthand" symbols are frequently used.

Symbol	Reading
$>$	is greater than
$<$	is less than
$\geqq$	equal to or greater than
$\leqq$	equal to or less than
$6'$	6 feet
$6''$	6 inches
Σ	the sum of
ΔL	change in L

1.10 NOMENCLATURE

Notation used in this book complies with that used in the 1989 ACI Code. The following list includes all of the notation used in this book and is compiled and adapted from a more extensive list given in Appendices B and C of the code.

A_c = area of concrete; gross area minus area of reinforcing

A_g = gross area of section ($A_c + A_s$)

A_s = area of reinforcing

A_s' = area of compressive reinforcement in a doubly reinforced section

A_v = area of shear reinforcing

A_1 = loading area in bearing

A_2 = gross area of bearing support member

E_c = modulus of elasticity of concrete

E_s = modulus of elasticity of steel

M = design moment

N = design axial load

V = design shear force

a = depth of equivalent rectangular stress block (strength design)

b = width of compression face of member

b_w = width of stem in a T-beam

c = distance from extreme compression fiber to the neutral axis (strength design)

d = effective depth, from extreme compression fiber to centroid of tensile reinforcing

e = eccentricity of a nonaxial load, from the centroid of the section to the point of application of the load

f_c = unit compressive stress in concrete

f_c' = specified compressive strength of concrete

f_s = stress in reinforcement

f_y = specified yield stress of steel

h = overall thickness of member; unbraced height of a wall

jd = length of internal moment arm

kd = distance from extreme compression fiber to the neutral axis (working stress)

n = modular ratio of elasticity: E_s/E_c

p = percent of reinforcing with working strength design, expressed as a ratio: A_s/A_g

s = spacing of stirrups

t = thickness of a solid slab

ρ = percent of reinforcing with ultimate strength design expressed as a ratio: A_s/A_g

ϕ = strength reduction factor (strength design)

2

MATERIALS AND PROPERTIES OF STRUCTURAL CONCRETE

This chapter presents discussions of the various basic ingredients of structural grade concrete and the factors that influence physical properties of the finished concrete. Other elements used to produce structural concrete are also discussed.

2.1 ORDINARY CONCRETE

Figure 2.1 shows the composition of ordinary concrete. The binder consists of the water and cement, whose chemical reaction results in the hardening of the mass. The binder is mixed with some aggregate (loose, inert particles) so that the binder coats the surfaces and fills the voids between the particles of the aggregate. For materials such as grout, plaster, and stucco, the aggregate consists of sand of reasonably fine grain size. For concrete the grain size is extended into the category of gravel, with the maximum particle size limited only by the size of the structure. The end product—the hardened concrete—is highly variable, due to choices for the individual basic ingredients, modifications in the

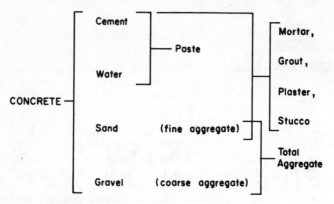

FIGURE 2.1 Composition of ordinary concrete.

mixing, handling, and curing processes, and possible addition of special ingredients.

2.2 CEMENT

The cement used most extensively in building construction is *portland cement*. Of the five types of standard portland cement generally available in the United States and for which the American Society for Testing and Materials has established specifications, two types account for most of the cement used in buildings. These are a general-purpose cement for use in concrete designed to reach its required strength in about 28 days, and a high-early-strength cement for use in concrete that attains its design strength in a period of a week or less.

All portland cements set and harden by reacting with water, and this hydration process is accompanied by generation of heat. In massive concrete structures such as dams, the resulting temperature rise of the materials becomes a critical factor in both design and construction, but the problem is not usually significant in building construction. A low-heat cement, is designed for use where the heat rise during hydration is a critical factor. It is, of course, essential that the cement actually used in construction corresponds to that employed in designing the mix, to produce the specified compressive strength of the concrete.

Air-entrained concrete is produced by using special cement and by introducing an additive during mixing of the concrete. In addition to improving workability (mobility of the wet mix), entrainment permits lower water–cement ratios and significantly improves the durability of the concrete. Air-entraining agents produce billions of microscopic air cells throughout the concrete mass. These minute voids prevent accumulation of water in cracks and other large voids which, on freezing, would permit the water to expand and result in spalling away of the exposed surface of the concrete.

2.3 MIXING WATER

Water must be reasonably clean, free of oil, organic matter, and any substances that may affect the actions of hardening, curing, or general finish quality of the concrete. In general, drinking-quality (potable) water is usually adequate. Salt-bearing seawater may be used for plain concrete (without reinforcing), but may cause corrosion of steel bars in reinforced concrete.

2.4 STONE AGGREGATE

The most common aggregates are sand, crushed stone, and pebbles. Particles smaller than $\frac{3}{8}$ in. in diameter constitute the *fine aggregate*. There should be only a very small amount of very fine materials, to allow for the free flow of the water–cement mixture between the aggregate particles. Material larger than $\frac{3}{8}$ in. is called the *coarse aggregate*. The maximum size of aggregate particle is limited by specification, based on the thickness of poured elements, spacing and cover of the reinforcing, and some consideration of finishing methods.

In general, the aggregate should be well graded, with some portion of large to small particles over a range to permit the smaller particles to fill the spaces between the larger ones. The volume of the concrete is thus virtually determined by the volume of the total aggregate, the water and cement going into the spaces remaining between the smallest aggregate particles. The weight of the concrete is determined largely by the weight of the coarse aggregate. Strength is also dependent to some degree on the integrity of the large aggregate particles.

2.5 ADMIXTURES

Substances added to concrete to improve its workability, accelerate its set, harden its surface, and increase its waterproof qualities are known as *admixtures*. The term embraces all materials other than the cement, water, and aggregates that are added just before or during mixing. Many of the proprietary compounds contain hydrated lime, calcium chloride, and kaolin. Calcium chloride is the most commonly used admixture for accelerating the set of concrete, but corrosion of steel reinforcement may be the consequence of its excessive use. Caution should be exercised in the use of admixtures, especially those of unknown composition.

Air-entrained concrete is produced by using an air-entraining portland cement or by introducing an air-entraining admixture as the concrete is mixed. In addition to improving workability, entrained air permits lower water–cement ratios (Sec. 2.6) and significantly improves the durability of hardened concrete. Air-entraining agents produce billions of microscopic air cells per cubic foot; they are distributed uniformly throughout the mass. These minute voids prevent the accumulation of water, which, on freezing, would expand and result in spalling of the exposed surface under frost action.

2.6 CONCRETE PROPERTIES

The primary index of strength of concrete is the specified compressive strength, designated f'_c. This is the unit compressive stress used for structural design and for a target for the mix design. It is usually given in units of psi, and it is common to refer to the structural quality of the concrete simply by calling it by this number: 3000-lb concrete, for example. For strength design, this value is used to represent the ultimate compressive strength of the concrete. For working stress design, allowable maximum stresses are based on this limit; specified as some fraction of f'_c.

Table 2.1 is reproduced from the 1963 ACI Code and indicates the various allowable stresses used in the working stress method in that code. The 1989 ACI Code (Ref. 1) contains vestiges of these standards in the alternate design method described in Appendix A. Use of these references is explained in the various portions of this part.

The value for the modulus of elasticity of concrete is established by a formula that incorporates variables of the weight (density) of the concrete and its strength. Distribution of stresses and strains in reinforced concrete is dependent on the concrete modulus, the steel modulus being a constant. This is discussed in Chapter 6.

When subjected to long-duration stress at a high level, concrete has a tendency to *creep*, a phenomenon in which strain increases over time under constant stress. This has effects on deflections and on the distributions of stresses between the concrete and reinforcing. Some of the implications of this for design are discussed in the chapters dealing with design of beams and columns.

Hardness of concrete refers essentially to its surface density. This is dependent primarily on the basic strength, as indicated by the value for compressive stress. However, surfaces may be somewhat softer than the central mass of concrete, owing to early drying at the surface. Some techniques are used to deliberately harden surfaces, especially those of the tops of slabs. Fine troweling will tend to draw a very cement-rich material to the surface, resulting in an enhanced density. Chemical hardeners can also be used, as well as sealing compounds that trap surface water.

The modulus of elasticity E_c of hardened concrete is a measure of its resistance to deformation. The magnitude of E_c depends on w, the weight of the concrete, and on f'_c, its strength. Its value may be determined from the expression $E_c = w^{1.5}33\sqrt{f'_c}$ for values of w between 90 and 155 lb/ft^3. For normal-weight concrete (145 lb/ft^3), E_c may be considered as equal to $57,000\sqrt{f'_c}$. [$E_c =$

TABLE 2.1   Allowable Stresses in Concrete

Description		For any strength of concrete in accordance with Section 502	For strength of concrete shown below			
			$f_c' =$ 2500 psi	$f_c' =$ 3000 psi	$f_c' =$ 4000 psi	$f_c' =$ 5000 psi
Modulus of elasticity ratio: n		29,000,000 $\overline{w^{1.5}\,33\sqrt{f_c'}}$				
For concrete weighing 145 lb per cu ft (see Section 1102)	n		10	9	8	7
Flexure: f_c						
Extreme fiber stress in compression	f_c	$0.45f_c'$	1125	1350	1800	2250
Extreme fiber stress in tension in plain concrete footings and walls	f_c	$1.6\sqrt{f_c'}$	80	88	102	113
Shear: v (as a measure of diagonal tension at a distance d from the face of the support)						
Beams with no web reinforcement*	v_c	$1.1\sqrt{f_c'}$	55*	60*	70*	78*
Joists with no web reinforcement	v_c	$1.2\sqrt{f_c'}$	61	66	77	86
Members with vertical or inclined web reinforcement or properly combined bent bars and vertical stirrups	v	$5\sqrt{f_c'}$	250	274	316	354
Slabs and footings (peripheral shear, Section 1207)*	v_c	$2\sqrt{f_c'}$	100*	110*	126*	141*
Bearing: f_c						
On full area			625	750	1000	1250
On one-third area or less†		$0.25f_c'$ $0.375f_c'$	938	1125	1500	1875

*For shear values for lightweight aggregate concrete see Section 1208.
†This increase shall be permitted only when the least distance between the edges of the loaded and unloaded areas is a minimum of one-fourth of the parallel side dimension of the loaded area. The allowable bearing stress on a reasonably concentric area greater than one-third but less than the full area shall be interpolated between the values given.

Source: Table 1002(a) from *Building Code Requirements for Reinforced Concrete* ACI 318-63; reproduced with permission of the publishers, American Concrete Institute.

$w^{1.5}0.043\sqrt{f'_c}$ for values of w between 1440 and 2480 kg/m³. For normal-weight concrete (2320 kg/m³), E_c may be considered as equal to 4730 $\sqrt{f'_c}$.]

In the design of reinforced concrete members, we employ the term n. This is the ratio of the modulus of elasticity of steel to that of concrete, or $n = E_s/E_c$. E_s is taken as 29,000 ksi [200,000 MPa].

Consider a concrete for which f'_c is 4000 psi and w is 145 lb/ft³. Then $E_c = 57,000 \sqrt{f'_c} = 57,000\sqrt{4000} = 3,600,000$ psi and $n = E_s/E_c = 29,000/3,600 = 8.055$. The values for n for four different strengths of concrete are given in Table 2.1 As is the usual practice, the values for n are rounded off to those given in the table.

As discussed in other sections, there are various controls that can be exercised to assure a desired type of material in the form of the hardened concrete. The three properties of greatest concern are the water content of the wet mix and the density and compressive strength of the hardened concrete. Design of the mix, handling of the wet mix, and curing of the concrete after casting are the general means for controlling the end product.

In addition to the basic structural properties, there are various properties of concrete that bear on its use as a construction material and in some cases on its structural integrity.

Workability. This term generally refers to the ability of the wet mixed concrete to be handled, placed in the forms, and finished while still fluid. A certain degree of workability is essential to the proper forming and finishing of the material. However, the fluid nature of the mix is largely determined by the amount of water present, and the easiest way to make it more workable is to add water. Up to a point this may be acceptable, but the extra water usually means less strength, greater porosity, and more shrinkage; all generally undesirable properties. Use of vibration, admixtures, and other techniques to facilitate handling without increasing the water content are often used to obtain the best-quality concrete.

Watertightness. It is usually desirable to have a generally nonporous concrete. This may be quite essential for walls or for floors consisting of paving slabs, but is good in general for protection of reinforcing from corrosion. Watertightness is obtained by having a well-mixed high-quality concrete (low water content, etc.), that is worked well into the forms and has dense surfaces with little cracking or voids. Subject to the continuous presence of water, however, concrete is absorptive and will become saturated. Moisture or waterproof barriers must be used where water penetration must be positively prevented.

Density. Concrete unit weight is essentially determined by the density of the coarse aggregate (ordinarily two-thirds or more of the total volume) and the amount of air in the mass of the finished concrete. With ordinary gravel aggregate and air limited to not more than 4% of the total volume, air dry concrete weighs around 145 lb/ft^3. Use of strong but lightweight aggregates can result in weight reduction to close to 100 lb/ft^3 with strengths generally competitive with that obtained with gravel. Lower densities are achieved by entraining air up to 20% of the volume and using very light aggregates, but strength and other properties are quickly reduced.

Fire Resistance. Concrete is noncombustible and its insulative, fire protection character is used to protect the steel reinforcing. However, under long exposure to fire, popping and cracking of the material will occur, resulting in actual structural collapse or a diminished capacity that requires replacement or repair after a fire. Design for fire resistance involves the following basic concerns:

1. *Thickness of Parts.* Thin slabs or walls will crack quickly, permitting penetration of fire or gases.
2. *Cover of Reinforcing.* More is required for higher fire rating of the construction.
3. *Character of the Aggregate.* Some are more vulnerable than others to fire actions.

Design specifications and building code regulations deal with these issues, some of which are discussed in the development of the building design illustrations in Chapter 12.

Shrinkage. Water-mixed materials, such as plaster, mortar, and concrete, tend to shrink in volume during the hardening process. For ordinary concrete, the shrinkage averages about 2% of the volume. The actual dimensional change of structural members is usually less, due to the presence of the steel bars; however, some consideration must be given to the shrinkage effects. Stresses caused by shrinkage are in some ways similar to those caused by thermal change; the combination resulting in specifications for minimum two-way reinforcing in walls and slabs. For the structure in general, shrinkage is usually dealt with by limiting the size of individual pours of concrete, as the major shrinkage ordinarily occurs quite rapidly in the fresh concrete. For special situations it is possible to modify the concrete with admixtures or special cements that cause a slight expansion to compensate for the normal shrinkage.

2.7 STEEL REINFORCEMENT

The steel used in reinforced concrete consists of round bars, mostly of the deformed type, with lugs or projections on their surfaces. The surface deformations help to develop a greater bond between the concrete and steel. The most common grades of reinforcing steel are grade 60 and grade 40, having yield strengths of 60,000 psi [414 MPa] and 40,000 psi [276 MPa], respectively. Properties for standard deformed reinforcing bars are given in a table on the inside back cover of this book, as considerable reference to them must be made in the computational work in the book.

Ample concrete protection, called *cover*, must be provided for the steel reinforcing. Cover is measured as the distance from the outside face of the concrete to the edge of a reinforcing bar. For reinforcement near surfaces not exposed to the ground or to weather, cover should be not less than ¾ in. [19 mm] for slabs, walls, and joists, and 1.5 in. [38 mm] for beams, girders, and columns. Where formed surfaces are exposed to earth or weather, the cover should be 1.5 in. [3.8 mm] for No. 5 bars and smaller and 2 in. [51 mm] for No. 6–No. 18 bars. For foundation construction poured directly against ground without forms, cover should be 3 in. [76 mm].

Where multiple bars are used in members (which is the common situation), there are both upper and lower limits for the spacing of the bars. Lower limits are intended to permit adequate development of the concrete-to-steel stress transfers and to facilitate the flow of the wet concrete during pouring. For columns, the minimum clear distance between bars is specified as 1.5 times the bar diameter or a minimum of 1.5 in. For other situations, the minimum is one bar diameter or a minimum of 1 in.

For walls and slabs, maximum center-to-center bar spacing is specified as three times the wall or slab thickness or a maximum of 18 in. This applies to reinforcement required for computed stresses. For reinforcement that is required for control of cracking due to shrinkage or temperature change, the maximum spacing is five times the wall or slab thickness or a maximum of 18 in.

For adequate placement of the concrete, the largest size of the coarse aggregate should be not greater than three-quarters of the clear distance between bars.

The essential purpose of steel reinforcing is to prevent the cracking of the concrete due to tension stresses. In the design of concrete structures, investigation is made for the anticipated structural actions that will produce tensile stress: primarily the actions of bending, shear, and torsion. However, tension can also be induced by shrinkage of the concrete during its drying out after the initial pour. Temperature variations may also induce tension in various situations. To provide for these latter actions, the ACI Code requires a minimum amount of reinforcing in members such as walls and slabs even when structural actions do not indicate any need. These requirements are discussed in the sections that deal with the design of these members.

In the design of most reinforced concrete members the amount of steel reinforcing required is determined from computations and represents the amount determined to be necessary to resist the required tensile force in the member. In various situations, however, there is a minimum amount of reinforcing that is desirable, which may on occasion exceed that determined by the computations. The ACI Code makes provisions for such minimum reinforcing in columns, beams, slabs, and walls. The minimum reinforcing may be specified as a minimum percentage of the member cross-sectional area, as a minimum number of bars, or as a minimum bar size. These requirements are discussed in the sections that deal with the design of the various types of members.

2.8 PRESTRESSING

Prestressing consists of the deliberate inducing of some internal stress condition in a structure prior to its sustaining of service loads. The purpose is to compensate in advance for some anticipated service load stress, which for concrete means some high level of tension stress. The "pre-" or "before" stress is therefore usually a compressive or reversal bending stress. This section discusses some uses of prestressing and some of the problems encountered in utilizing it for building structures.

Use of Prestressing

The principal use of prestressing is for spanning elements, in which the major stress conditions to be counteracted are tension due to bending and diagonal tension due to shear. A principal advantage of prestressing is that when properly achieved, it does not result in the natural tension cracking associated with ordinary reinforced concrete. Since flexural cracking is proportionate to the depth of the member, which in turn is proportionate to the span, the use of prestressing frees spanning concrete members from the span limits associated with ordinary reinforcing. Thus gigantic beam cross sections and phenomenal spans are possible—and indeed have been achieved, although mostly in bridge construction.

The cracking problem also limits the effective use of very high concrete strengths with ordinary reinforcing. Free of this limit, the prestressed structure can utilize effectively the highest strengths of concrete achievable, and thus weight saving is possible, resulting in a span-to-weight ratio that partly overcomes the usual massiveness of spanning concrete structures.

The advantages just described have their greatest benefit in the development of long, flat-spanning roof structures. Thus a major use of prestressing has been in the development of precast, pre-stressed units for roof structures. The hollow-cored slab, single-tee, and double-tee sections shown in Fig. 2.2 are the most common forms of such units—now a standard part of our structural inventory. These units can also be used for floor structures, having a major advantage when span requirements are at the upper limits of feasibility for ordinary reinforced construction.

Some other uses of prestressing in building structures are the following:

1. *For Columns.* Concrete shafts may be prestressed for use as building columns, precast piles, or posts for street lights or signs. In this case the prestressing compensates for bending, shear, and torsion associated with service use and handling during production, transportation, and installation. The ability to use exceptionally high strength concrete is often quite significant in these applications.

2. *For Two-Way Spanning Slabs.* Two-way, continuous pre-stressing can be used to provide for the complex deforma-

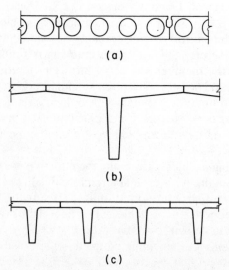

(a)

(b)

(c)

FIGURE 2.2   Forms of typical precast concrete spanning units.

tions and stress conditions in concrete slabs with two-way spanning actions. A special usage is that for a paving slab designed as a spanning structure where ground settlement is anticipated. Crack reduction may be a significant advantage in these applications.

3. *Tiedown Anchors.* Where exceptionally high anchorage forces must be developed, and development of ordinary tension reinforcing may be difficult or impossible, it is sometimes possible to use the tension strands employed for prestressing. Large abutments, counterforts for large retaining walls, and other elements requiring considerable tension anchorage are sometimes built as prestressed elements.

4. *Horizontal Ties.* Single-span arches and rigid frames that develop outward thrusts on their supports are sometimes tied with prestressing strands.

For any structure it is necessary to consider various loading conditions that occur during construction and over a lifetime of use. For the prestressed structure this is a quite complex issue, and design must incorporate many different events over the life of the structure. For common usages, experience has produced various empirical adjustments (educated fudge factors) that account for the usual occurrences. For unique applications there must be some reasonable tolerance for errors in assumptions or some provision for tuning up the finished structure. The prestressed structure is a complex object, and design of other than very routine elements should be done by persons with considerable training and experience.

Pretensioned Structures

Prestressing is generally achieved by stretching high-strength steel strands (bunched wires) inside the concrete element. The stretching force is eventually transferred to the concrete, producing the desired compression in the concrete. There are two common procedures for achieving the stretching of the strands: pretensioning and post-tensioning.

Pretensioning consists of stretching the strands prior to pouring the concrete. The strands are left exposed and as the concrete hardens it bonds to the strands. When the concrete is sufficiently hardened, the external stretching force is released and the strand tension is transferred to the concrete through the bond action on the strand surfaces. This procedure requires some substantial element to develop the necessary resistance to the jacking force used to stretch the strands before the concrete is poured. Pretensioning is used mostly for factory precast units, for which the stretch force resisting element is the casting form, sturdily built of steel and designed for continuous, multiple use.

Pretensioning is done primarily for cost-saving reasons. There is one particular disadvantage to pretensioning: it does not allow for any adjustment and the precise stress and deformation conditions of the finished product are only approximately predictable. The exact amount of the strand bonding and the exact properties of the finished concrete are somewhat variable. Good quality control in production can keep the range of variability within some bounds, but the lack of precision must be allowed for in the design and construction. A particular problem is that of control of deflection of adjacent units in systems consisting of side-by-side units.

Post-Tensioned Structures

In post-tensioning the prestressing strands are installed in a slack condition, typically wrapped with a loose sleeve or conduit. The concrete is poured and allowed to harden around the sleeves and the end anchorage devices for the strand. When the concrete has attained sufficient strength, the strand is anchored at one end and stretched at the other end by jacking against the concrete. When the calibrated jacking force is observed to be sufficient, the jacked end of the strand is locked into the anchorage device, pressurized grout is injected to bond the strand inside the sleeve, and the jack is released.

Post-tensioning is generally used for elements that are cast in place, since the forms need not resist the jacking forces. However, it may also be used for precast elements when jacking forces are considerable and/or a higher control of the net existing force is desired.

Until the strands are grouted inside the sleeves, they may be rejacked to a higher stretching force condition repeatedly. In some situations this is done as the construction proceeds, permitting the structure to be adjusted to changing load conditions.

Post-tensioning is usually more difficult and more costly, but there are some situations where it is the only alternative for achieving the prestressed structure.

2.9 CONCRETE MIX DESIGN

From a structural design point of view, mix design basically means dealing with the considerations involved in achieving a particular design strength, as measured by the value of the fundamental property: f'_c. For this consideration alone, principal factors are the following:

1. *Cement Content.* The amount of cement per unit volume is a major factor determining the richness of the cement–water paste and its ability to fully coat all of the aggregate particles and fill the voids between them. Cement content is normally measured in terms of the number of sacks of cement (1 cubic foot each) per cubic yard of concrete mixed. The average for structural concrete is about 5 sacks per yard. If tests show that the mix exceeds or falls short of the desired results, the cement content is increased or decreased. The cement is by far the costliest ingredient, so its volume is critical in cost control.

2. *Water–Cement Ratio.* This is expressed in terms of gallons of water per sack of cement or gallons of water per yard of mixed concrete. The latter is usually held very close to an average of about 35 gallons per yard; less and workability is questionable; more and strength becomes difficult to obtain. Attaining of very strong concrete usually means employing various means to reduce water content and improve the ratio of cement to water without losing workability.

3. *Fitness Modulus of the Sand.* If too coarse, the wet mix will be grainy and finishing of surfaces will be difficult. If too fine, an excess of water will be required, resulting in high shrinkage and loss of strength. Grain size is controlled by specification.

4. *Character of the Coarse Aggregate.* Shape, size limits, and type of material must be considered. Since this represents the major portion of the concrete volume, its properties are quite important to strength, weight, fire performance, and so on.

Concrete is obtained primarily from plants that mix the materials and deliver them in mixer-trucks. The mix design is developed cooperatively with the structural designer and the management of the mixing plant. Local materials must be used, and experience with them is an important consideration.

2.10 SPECIAL CONCRETES

Within the range of the general material discussed here, there are many special forms of concrete used for special situations and applications. Some of the principal ones are the following.

Lightweight Structural Concrete. This is concrete that achieves a significant reduction in weight while retaining sufficient levels of structural properties to remain feasible for major structural usages. Maximum weight reduction is usually in the range of 30%. Strength levels may be kept reasonably high, but some loss in stiffness (modulus of elasticity) is inevitable, so deflections become more critical. The principal means for achieving weight reduction is by use of other than stone for the coarse aggregate. Some natural materials may be used for this, but the more typical uses are with synthetic materials. One major use is for the concrete fill on top of formed sheet steel decking in steel-framed structures.

Super Heavyweight Concrete. For some purposes, it may be desired to achieve an increase in the concrete density (unit weight). A simple means for achieving this is to use a particularly heavy material for the coarse aggregate. Some of the heaviest natural materials are metal ores, but careful analysis must be made of their interactions with the chemical processes in the concrete.

Insulating Concrete. Use of super lightweight aggregates, usually natural or "popped" mineral materials, together with deliberately entrapped air (foaming), can produce concrete with densities below 30 lb/ft^3. Compressive strength drops to a few hundred psi, so major structural usage is out of the question, but the material is used for fill on top of roof decks and for some situations to insulate steel framing from fire.

Super-Strength Concrete. Through the use of specially selected materials, addition of water-reducing and density-enhancing admixtures, and very special mixing, handling, and curing, concrete strengths in the range of 20,000 psi can now be achieved. Major usage to date has been for the lower structures of very tall concrete buildings. This requires major effort and considerable expertise and is very expensive, but is now quite routinely accomplished where it offers significant value. The nature of this material is out of the range of our traditional procedures and specifications, so its design control is still being developed.

Fiber-Reinforced Concrete. Experiments have been conducted over many years with the inclusion of fibrous elements in the concrete mix with the intention of giving an enhanced tension resistance to the basic hardened concrete. Steel needles, glass, and various mineral fibers have been used. The resulting tensile-enhanced material tends to resist cracking, permit very thin, flexible elements, resist freezing, and permit some applications without steel reinforcing rods. Only minor structural applications have been attempted, but the material is now commonly used for pavements and for thin roof tiles and cladding panels.

As materials research intensifies—both commercially and with some nonprofit sponsorship—new materials that offer new potential uses are sure to increase in number. Still, traditional sand-and-gravel concrete remains in wide use for common situations.

2.11 CRACKING

All rigid, tensile-weak materials are highly susceptible to some form of cracking. This is a major concern for concrete, as well as for stone, masonry, and plaster and building construction uses. The major sources of cracking in concrete are the following.

Shrinkage. The water-mixed material shrinks in volume when it dries. In ordinary concrete, this is typically about a 2 to 3% volume change. Long elements, thin elements, and continuous elements with many geometric discontinuities are most susceptible to shrinkage cracking. As concrete building structures ordinarily have all of these, some design considerations must usually be made.

Temperature Change. Thermal expansion and contraction is similar to shrinkage in terms of the effects of volume change. This is of particular concern in regions with a very wide range of outdoor temperatures and applies especially to structures exposed to the weather.

Structural Action. Any structural action other than simple direct compression will produce some tensile stress in the material of a structure. Where major bending or torsion occurs, this is usually the most severe. In ordinary reinforced concrete beams and spanning slabs some cracking of the concrete in the vicinity of the tensile reinforcement is inevitable. (See discussions in Secs. 2.8 and 6.3.)

Settlement of Supports. Concrete structures are usually quite heavy and are consequently vulnerable to settlement on soils in which this offers a problem. Add this to the usual rigid nature of concrete structures and the potential for problems is considerable.

A major means for reduction of cracking is the placing of steel reinforcement in the necessary amounts and in strategic locations. This is basically what design of reinforced concrete is all about! However, the problems resulting in cracking may also be reduced by intelligent planning and detailing of the construction, selection of proper concrete mixes, good curing procedures, and other generally good design and construction practices. In fact, using more reinforcing materials should be the last resort, as steel reinforcement is very expensive.

Use of prestressing or inclusion of fibrous material in mixes may also be done with a major intention of reducing cracking. In many situations some minor cracking does not reduce the load-carrying capacity of concrete structures. Cracking is more frequently a concern for appearance, for water penetration, or for other nonstructural reasons. Where the latter is the case, compensations may be achieved through applied finishes or other means, rather than striving to totally eliminate cracking of the concrete.

3

CONSIDERATIONS FOR PRODUCTION OF CONCRETE

3.1 GENERAL CONCERNS

Because concrete is a mixture in which a paste made of portland cement and water binds together fine and coarse particles of inert materials, known as aggregates, it is readily seen that by varying the proportions of the ingredients innumerable combinations are possible. These combinations result in concrete of different qualities. When the cement has hydrated, the plastic mass changes to a material resembling stone. This period of hardening is called *curing*, in which three things are required: time, favorable temperatures and the continued presence of water.

To fulfill requirements it is essential that the hardened concrete have, above all else, *strength* and *durability*. In order that the concrete in its plastic form may be readily placed in the forms, another essential quality is *workability*. When watertightness is required, concrete must be *dense* and *uniform* in quality. Hence it is seen that in determining the various proportions of the mixture the designer must have in mind the purpose for which the concrete is to be used and the exposure to which it will be subjected.

The following factors regulate the quality of the concrete: suitable materials, correct proportions, proper methods of mixing and placing, and adequate protection during curing.

3.2   CONCERNS FOR STRUCTURAL CONCRETE

For the production of major structural elements a principal concern is the specific value of the compressive strength of the finished concrete. This value is assumed in the design work, striven for in the design specifications, mixing, handling, curing, and general production of the concrete, and if not achieved, can cause major problems with regard to certifying the safety of the structure. Major efforts are made by various parties—designers, concrete producers, builders, testing labs, and code-enforcing agencies—to assure the proper quality of the finished concrete.

3.3   CAST-IN-PLACE CONCRETE

Most concrete for buildings is cast in place; that is, the wet mix is deposited and formed at the place where the finished concrete is desired. This is now generally referred to as *sitecast concrete*, since the location is usually at a building site. This is compared to *precast concrete*, which refers to the process of casting elements and then moving them to the place they are to be used.

Concrete for sitecast construction is typically brought to the site by the familiar concrete-transporting mixer trucks, with the large rotating barrels. The mix is prepared at a central batching plant, where controls of the materials may be carefully monitored. However, the transporting to the site, proper mixing in the truck, discharging from the truck and depositing in the forms, and handling for placement, finishing, and curing are all subject to the level of responsibility and craft exercised by the people involved. Site conditions in terms of accessibility and weather can be highly critical to the work, requiring extreme measures in some situations to control all the stages in the production process. This book is not about construction processes or their management, but some awareness of the issues and limitations is helpful in developing reasonable designs for concrete structures.

3.4 FORMING

An inherent property of concrete is that it may be made in any shape. The wet mixture is placed in *forms* constructed of wood, metal, or other suitable material in which it hardens or sets. The forms must be put together with quality workmanship, holding to close dimensional tolerances. Formwork should be strong enough to support the weight of the concrete and rigid enough to maintain position and shape. In addition, formwork should be tight enough to prevent the seepage of water and designed to permit ready removal.

Timber used for forms is usually surfaced on the side that comes in contact with the concrete, and frequently is oiled or otherwise sealed. This fills the pores of the wood, reduces absorption of water from the concrete mixture, produces smoother concrete surfaces, and permits the form boards to be more easily removed.

Steel forms have the decided advantage of being more substantial if they are to be reused. Steel gives smoother surfaces to the concrete, although it is almost impossible to avoid showing the joints. For ribbed floors, metal pans and domes are used extensively, and columns, circular in cross section, are invariably made with metal forms.

Because the formwork for a concrete structure constitutes a considerable item in the cost of the completed structure, particular care should be exercised in its design. It is desirable to maintain a repetition of identical units so that the forms may be removed and reused at other locations with a minimum amount of labor.

There are no exact rules concerning the length of time the forms should remain in place. Obviously they should not be removed until the concrete is strong enough to support its own weight in addition to any loads that may be placed on it. Also, too early removal of forms introduces the possibility of excessive deflections. Sometimes the side forms of beams are removed before the bottom forms. When this is done, posts or shoring are placed under the bottoms of the members to give additional support. This is called *reposting* or *reshoring*. The minimum period during which forms must remain in place before stripping is usually governed by the local building code.

3.5 PLACING AND FINISHING

As soon as cement is mixed with water a chemical action begins that eventually results in the hardening of the concrete. The first stage of this is the wet mix, which has the character of a thick, viscous fluid. In a short time, however, an initial hardening (called *set*) occurs, and the fluid nature of the mix fades. Before the initial set occurs, the concrete must be fully placed in the forms—a matter of only a few hours with ordinary mixes. Add up the time for loading the trucks at the batching plant, driving the trucks to the site, emptying the trucks and moving the mixed concrete to its desired location, and handling it to fill the forms, and there is not much time for leisure in the operation.

Despite the haste required, the wet concrete must be carefully handled so as not to cause segregation, which consists of the partial separation of the ingredients. This can occur if the concrete sits idly in its wet state, is dropped too far when deposited, or is moved around too much in the forms before settling into place. This is a situation for careful supervision, but is mostly up to the skills and care of the workers involved.

Surfaces of the concrete in contact with forms will derive a primary form and finish from the surfaces of the forms. Unformed surfaces (usually the top) may receive various treatments, the simplest being a simple struck surface, produced by smoothing with a board or rough wood tool. This is essentially an unfinished surface, which may be additionally treated during the initial hardening or later.

Selection of forming materials may be made to achieve certain desired form or finish of the cast concrete. Of particular concern is the surface of forms. The smoothest, cleanest surfaces are achieved with steel, fiberglass, or plastic-coated plywood forms. Special surfaces may be achieved with form-lining elements or by coating the forms with various materials. One special material of the latter type is a retarder which slows down the hardening of the surface cement–water paste, so that the face of the concrete is partly stripped away when the forms are removed, revealing the embedded aggregate just below the surface. Sand-blasting of the freshly cast surface—possibly in combination with use of a retarder—can also produce a roughened, aggregate-exposed surface.

Many other finishes can be achieved, in the process of mixing, initial forming and finishing, or reworking of the hardened surfaces. For structural members, concern must be given to the degree of surface loss that occurs, which may affect loss of member cross sections and concrete cover for the steel reinforcement.

Other concerns for exposed surfaces include the possible effects of various elements used in the construction, of which some examples are the following.

Bar Chairs and Spacers. These are used to hold steel bars above the bottom of the forms. They must sit on the forms, and thus their feet will be at the surface when forms are removed. Plastic elements or steel with plastic coating should be used where exposure may cause rust spotting on surfaces.

Form Ties. Walls are typically formed with two surfaces that are tied to each other across the void (concrete-filled) space to prevent their bowing out from the hydraulic pressure of the wet concrete. These ties become embedded and their outer ends are hard to conceal.

Joints in Forms. Large concrete members—long beams, walls, undersides of slabs—must be formed of many units with joints between units. Concrete exposed to view will show these joints, unless the surfaces are completely reworked by some process.

Except for pavements, much structural concrete occurs in basements or foundations, or is covered by other construction, so that surface finishes are often of little concern, other than that they should be reasonably dimensionally true. Dimensional accuracy and reasonable sturdiness may be the only required attributes of forming and simple struck surfaces may be adequate for nonformed faces of the concrete.

3.6 CURING AND PROTECTION

Regardless of the care taken in proportioning, mixing, and placing, first-quality concrete can be obtained only when due consideration and provision are made for curing. The hardening of concrete is due to the chemical reaction between the water and cement. This hardening continues indefinitely as long as moisture is present and the temperatures are favorable. The initial set does not begin until two or three hours after the concrete has been mixed. During this interval moisture evaporates, particularly on the exposed surfaces, and unless provision is made to prevent the loss of moisture, the concrete will craze. A typical specification requires that the concrete be so protected that there is no loss of moisture from the surface for a period of seven days when normal portland cement is used and three days when the cement is of high early strength.

To prevent the loss of moisture during curing several methods may be employed. When hard enough to walk on, slabs may be covered with burlap which is kept wet or with a suitable building paper with the edges pasted down. Another method is to cover the slabs with a 1-in. layer of wet sand or sawdust. Another method sometimes resorted to is the continuous sprinkling of the exposed surfaces with water. The early removal of forms permits undue evaporation; hence the forms should be allowed to remain for as long a period as is practicable. In addition to strength and durability, controlled curing is one of the best precautions in making a watertight concrete.

The period of protection against evaporation of moisture varies with the type of structure and climatic conditions. Thin sections of concrete placed during hot weather require an increased period of protection.

Low temperatures during the period of curing produce concrete of lower strength than concrete cured at 72°F. Freezing of concrete before it has cured should never be permitted to occur, for the resulting concrete will be of poor quality and indeterminate strength.

Although special precautions are required, concrete work may be continued during severe weather conditions. To keep the concrete above freezing the materials may be heated before mixing or the concrete may be protected with suitable covers or kept in heated enclosures. If the weather is only moderately cold, heating the water used for mixing may be a sufficient precaution. In more severe weather, it may be necessary to heat both water and aggregates. The materials should never have a temperature exceeding 90°F when deposited.

One common method of protecting concrete is to cover it with a thick layer of straw and tarpaulins. Canvas enclosures heated by steam give excellent protection, since desirable temperatures may be maintained and the concrete is protected against drying out. If other heating devices are used, care should be exercised to see that moisture is not evaporated from the concrete. Most structural members in reinforced concrete buildings are not exposed to the weather but exterior columns and spandrel beams frequently are. Alternate cycles of freezing and thawing, wetting and drying, and prolonged periods of surface wetting diminish the durability of concrete. Consequently, the degree of exposure must be taken into account when designing the mix.

In climates in which freezing occurs, entrained air should be used in all exposed concrete. Sulfate-resisting cements should be used where concrete is exposed to seawater or comes in contact with sulfate-bearing soils. When conditions of exposure are severe, a low water–cement ratio should be used even though strength requirements may be met with a higher value.

3.7 DESIGN AND PRODUCTION CONTROLS

Structural designers ordinarily document their work in the form of a set of written computations. These computations will include a listing of design criteria: codes and standards used, concrete strength and steel type used, design loadings assumed, and so on. The computations are concluded by a listing of the design decision information: required shape and dimensions of concrete elements and the positions, number, and size of reinforcing bars. In most cases some sketches are used in the computations, to indicate the arrangement of reinforcing in member cross sections and the locations of bar cutoffs, extensions, bend points, and so on.

Structural computations are not ordinarily used to transmit information to the builder. For this purpose, it is ordinarily the practice to produce a set of contract documents: working drawings and specifications. Translation of the computations into the construction documents is normally done in the design office, and the designer should understand this process and be able to check the final form of the construction documents in order to assure that his or her design has been properly translated.

While the construction documents (if thoroughly executed) completely delineate the finished structure, the builders must usually produce a second set of documents that explain more directly to the work force how to make the concrete forms, provide falsework (the supports required during pouring and curing of the concrete), and fabricate and install the steel bars. Although the correctness of this translation is the responsibility of the builder, the designer should also verify the accuracy to avoid mistakes that will delay the construction.

At various stages of the construction, the adequacy of the work should be verified by the designer. The proper shape, details, and dimensions of forms, and the proper installation of reinforcing should be checked prior to pouring of the concrete. Installation in the forms of inserts for attachment, piping, electrical conduit, blocking for ducts, wiring, and piping chases, and so on, should be inspected to assure that they do not critically reduce the structural capacity of affected members.

The quality of the finished concrete will be affected by many factors, as discussed later in this chapter. Control of these factors by the designer is done primarily by well-written specifications, but a little nagging during construction doesn't hurt.

3.8 INSPECTION AND TESTING

If the operation is of sufficient magnitude, concrete made of various proportions and with aggregates from the sources proposed for use on the job should be tested before construction of the building is started. In the usual procedure several combinations are tested by using at least four different water–cement ratios. The results of the tests are then plotted and the most economical mix that will produce a concrete of the desired strength, density, and workability is chosen. It is customary to continue testing the concrete during construction, particularly if there are changing weather conditions or if a change is made in the sources from which the aggregates are obtained.

The two most common tests of concrete are the slump test for determining the degree of workability of the fresh concrete and the compression test on cylinders of cured concrete to establish its strength. The effectiveness of these tests in quality control of concrete production depends on obtaining truly representative samples of fresh concrete and following standard procedures during testing. The American Society for Testing and Materials issues ASTM Standards covering sampling and testing, which are prescribed procedures under the ACI Code.

Slump Test

Although the terms *consistency* and *workability* are not strictly synonymous, they are closely related. Consistency may be loosely defined as the wetness of the concrete mixture; it is an index of the ease with which concrete will flow during placement. A concrete is said to be workable if it is readily placed in the forms for which it is intended; for instance, a concrete of given consistency may be workable in large open forms but not in small forms containing numerous reinforcing bars. With this understanding, the slump test may be considered a measure of the workability of fresh concrete.

The equipment for making a slump test consists of a sheet metal truncated cone 12 in. high with a base diameter of 8 in. and a top diameter of 4 in. Both top and bottom are open. Handles are attached to the outside of the mold. When a test is made, freshly mixed concrete is placed in the mold in a stipulated number of layers and each is rodded separately a specified number of times with a steel rod. When the mold is filled, the top is leveled off and the mold lifted at once. The slumping action of the concrete is measured by taking the difference in height between the top of the mold and the top of the slumped mass of concrete (Fig. 3.1).

If the concrete settles 3 in., we say that the particular sample has a 3-in. slump. Thus the degree of consistency of the concrete is ascertained. Recommended slump ranges for various types of construction are given in construction standards.

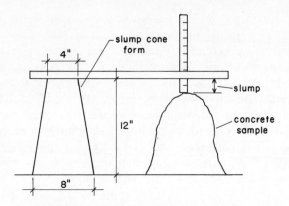

FIGURE 3.1 Slump test for wet concrete.

Compression Test

Tests of compressive strength are made at periods of 7 and 28 days on specimens prepared and cured in accordance with prescribed ASTM testing procedures. The specimen to be tested is cylindrical in shape and has a height twice its diameter. The standard cylinder is 6 in. in diameter and 12 in. high when the maximum size of the coarse aggregate does not exceed 2 in. For larger aggregates, the cylinder should have a diameter at least three times the maximum size of the aggregate and its height should be twice the diameter.

The mold used for the cylinders is made from metal or other nonabsorbent material such as paraffined cardboard. It is placed on a smooth plane surface (glass or metal plate) and filled with freshly made concrete in a specified number of layers. Each layer is consolidated by rodding or vibrating, either method being acceptable for concretes with a slump of 3 to 1 in. If the slump is greater than 3 in., the concrete must be rodded; if it is less than 1 in., it must be vibrated. As soon as casting of the cylinder is complete, the top of the specimen is covered to prevent the concrete from drying.

Because the strength of a specimen is greatly affected by temperature changes, exposure to drying, and disturbances due to movement, it is customary to keep it at the site of operation for 24 hours. It is then taken to the laboratory and cured under controlled conditions in accordance with standard ASTM procedures. At the end of the curing period, each specimen is placed in the testing machine and a gradually increasing compressive load is applied until the specimen fails. The load causing failure is recorded, and this load, divided by the cross-sectional area of the cylinder, gives the ultimate compressive unit stress of the specimen. The same test is made on other specimens taken at the same time and cured under similar conditions, which, of course, results in a range of values for the compressive strength.

3.9 INSTALLATION OF REINFORCEMENT

To facilitate the shop fabrication and field installation of reinforcing bars, the bar supplier usually prepares a set of drawings—commonly called the *shop drawings*. These drawings consist of the supplier's interpretation of the engineering contract drawings, with the information necessary for the workers who fabricate the bars in the shop and those who install the bars in the field prior to pouring of the concrete. The exact cut lengths of bars, the location of all bends, the number of each type of bar, and so forth, will be indicated on these drawings. While the correctness of these drawings is the responsibility of the supplier, it is usually a good idea for the designer to verify the drawings in order to reduce mistakes in the construction.

Reinforcing bars must be held firmly in place during the pouring of the concrete. Horizontal bars must be held up above the forms; vertical bars must be braced from swaying against the forms. The positioning and holding of bars is done through the use of various accessories and a lot of light-gage tie wire. When concrete surfaces are to be exposed to view after being poured, it behooves the designer to be aware of the various problems of holding bars and bracing forms, since many of the accessories used ordinarily will be partly in view on the surface of the finished concrete.

Installation of reinforcing may be relatively simple and easy to achieve, as in the case of a simple footing or a single beam. In other cases, where the reinforcing is extensive or complex, the problems of installation may require consideration during the design of the members. When beams intersect each other, or when beams intersect columns, the extended bars from the separate members must pass each other at the joint. Consideration of the "traffic" of the intersecting bars at such joints may affect the positioning of bars in the individual members.

3.10   PRECAST CONCRETE

Precasting refers to the process of construction in which a concrete element is cast somewhere other than where it is to be used. The other place may be somewhere else on the building site or away from the site, probably in a casting yard or factory. The precast element may be prestressed, may be of ordinary reinforced construction, or may even be without reinforcement. The single precast element may be a component of a general precast concrete system, or may serve a singular purpose in a construction system of mixed materials or types of elements. This section gives some discussion of the uses of precasting and the problems encountered in designing precast elements and systems.

Use of Precasting

The technique of precasting is utilized in a variety of ways. Undoubtedly, the most widely used precast element is the ordinary concrete block—called a CMU (concrete masonry unit). Most structural masonry is made from these units. In ordinary construction the block form shown in Fig. 3.2a is commonly used; for reinforced masonry construction (used exclusively in zones of high seismic risk) a different form is used. Construction with CMUs is discussed in Sec. 3.11.

Another widely used precast element is the tilt-up wall unit, shown in Fig. 3.2b. This element is sitecast in a horizontal position, then tilted up and moved by a crane to its desired location. The casting bed usually consists of the building floor slab on grade, resulting in a considerable reduction in forming cost. This type of construction is widely used for one-story and low-rise

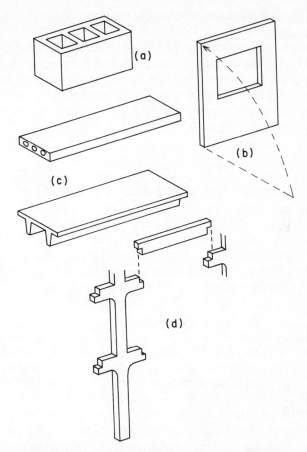

FIGURE 3.2   Structural elements of precast concrete.

commercial structures in the southern and western regions of the United States.

As discussed in Chapter 12, some of the most widely used prestressed elements are the flat-spanning units of hollow core or tee form used for roof and floor construction. These units are produced in casting factories in continuous production processes.

Structural systems consisting of connected components of precast concrete have been produced in great variety. Some of these have been produced as patented, manufactured systems, but mostly they have been the single, innovative products of individual designers.

Design and construction of precast concrete is strongly influenced by the standards and publications of the Precast Concrete Institute (PCI). Industry products are largely developed in conformance with these standards. Anyone contemplating the design of a unique system or element of precast concrete is advised to investigate the information available from PCI.

Advantages of Precast Concrete

There are various reasons for considering the use of precast concrete construction. In some cases the choice is between precasting and ordinary construction of cast-in-place concrete, with elements formed and cast at the location where they are to be used. In other cases the choice may be between using precast elements or some other material or type of construction. The following are some advantages offered by the precasting process, generally in comparison to cast-in-place construction.

Faster Site Work.   Cast-in-place concrete construction usually proceeds quite slowly, requiring construction of forms, installation of reinforcing, pouring of concrete, hardening to sufficient strength to permit removal of forms, and so on. Erection of precast elements is more akin to construction with steel or timber structures, and the faster site work may be an advantage where construction time is highly constrained. However, it is the total building construction time that is significant, not just the time to get up the structure. If time cannot also be gained in other parallel construction activities, the rapidly erected structure may just sit there and wait for the other project work to catch up.

Forming Economies.   For the ordinary cast-in-place concrete structure, a major portion of the total cost is represented by the forming. This includes the cost of the construction, bracing and support, and removal of the forming. Some reuse of items may be possible, but the process tends to use materials up rapidly. Precasting offers more potential for reuse of forms, even with sitecast construction. Factory processes involve extensive reuse of forms or production by forming processes such as extrusion. Reduction of on-site labor costs may often be the major gain in this area.

Quality Control. Precision of detail, quality and uniformity of finishes, and uniformity of concrete properties (color, density, compaction, etc.) may be assured in factory production to a degree not possible with site casting. Here it is not just precasting but factory conditions that are the issue. All of this is more true if the element is a standard manufactured product subject to ongoing quality control in its production. This is obviously of greatest concern for construction elements exposed to view, especially wall components.

Use of Predesigned Elements and Systems. Face it—design effort means time and money. Use of a standard predesigned building construction component means a shortcut in design development effort. If the component is part of a system, with system-wide concerns for total building utilization carefully preconsidered and standardized, the savings in design effort and the relative assurance of end results may be quite significant. This is treacherous ground for designers, with the possible end result being a direct connection between supplier and building owner, leaving the designer out in the cold.

Utilization of High-Quality Concrete and Prestressing. The potential for utilization of the structural properties of very high strength concrete usually occurs in association with prestressing (versus ordinary reinforcing). Factory-produced concrete is routinely of higher quality than the concrete produced by site casting. In some situations, denser surfaces, lower permeability, and reduced shrinkage may also be significant. This may be a factor in the use of mixed systems of precast and case-in-place elements.

Problems with Precast Concrete

As with any form of construction, there are some particular problems associated with precast concrete. These are not necessarily insurmountable, but designers should be aware of them and of the considerations that may be required in using the construction method. The following are some major concerns that may be of significance in various situations.

Handling and Transporting. Concrete construction is usually heavy—precast elements included. Precast units are usually of considerable size, and the combination presents a major problem of handling and transporting the heavy and relatively fragile units. Stresses induced during handling and erecting of units may be significant structural design concerns. Use of factory-cast units is usually feasible only within some reasonable distance from the factory.

Cost. The cost of production, handling, and transporting of precast units is considerable. There needs to be some considerable list of other advantages to make this form of construction generally competitive—not so much with alternative cast-in-place concrete, but with other materials and systems. In the end, the total building construction cost is most significant, not just structural cost.

Connections. From a general construction development point of view, the single biggest problem in design with precast elements is usually the connection of elements. From a structural response consideration, it is here that the major difference occurs between precast and ordinary cast-in-place concrete construction. The completely precast structure has more in common with structures of steel and timber than with ordinary concrete structures. The adequate development of individual connections is a problem, but almost more significant is the overall loss of natural continuity and inherent stability of the fully cast-in-place system. A major effort by the PCI is in the development of recommended practices for design and construction of connections of precast elements. Connection considerations may be a major factor in a decision to use some of the techniques discussed in this section.

Integration. This refers to the general problem of incorporating the precast concrete elements into the general building construction. A major problem to be dealt with in this regard is the loss of some opportunities that are present with other types of construction. Installation of hidden items such as wiring, piping, ducts, and housings for light switches, power outlets, recessed lighting fixtures, bathroom medicine cabinets, fire hose cabinets, and exit signs is made somewhat more difficult. With light wood frames, ordinary cast-in-place concrete, and most other types of construction, precise locations of these items and the actual installation can be done during site construction work. With precast concrete construction, provisions must be made in advance—a procedure that is not impossible, but that does not fit with the routine operations with which most designers and builders are familiar. This is really a minor matter of adjustment of procedures and scheduling, but one that can create difficulties if it is not anticipated and provided for.

Seismic-Effects. Usage of precast concrete has received some setbacks in response to the performance of some structures during major seismic events. Promotional literature of the wood and steel industry frequently contain some dramatic pictures of collapsed structures of precast concrete. This is indeed a major problem to be dealt with in design, especially of connections of structural components. The loss of continuity in changing from cast-in-place to precast construction is a major concern. Learning from failures (a basic process in all structural design areas), the precast industry has developed more stringent criteria and techniques. Nevertheless, the seismic response of heavy, individually stiff and brittle elements, with many joints in the assembled system, makes the precast structural system less than ideal in seismic response. In spite of this, some systems—such as the tilt-up wall—see ongoing extensive use in regions of high seismic risk, attesting to the fact that design can be effectively achieved when other factors are sufficiently persuasive in the overall decision of system selection.

Mixed Systems: Sitecast and Precast

The completely precast concrete structure does not represent the widest usage of precasting. Precast components are used in many situations in conjunction with a building structure and general construction with other materials and systems. This is, of course, the *usual* situation; few buildings are all steel, all wood, all concrete, or all masonry.

Precast concrete decks, especially of the hollow-cored form, are used with frames of steel or sitecast concrete and with bearing walls of masonry or poured concrete. Precast concrete wall panels are used with structures of steel, wood, and sitecast concrete. The separate component/separate material situation is a common one in building construction. Of course, the individual functions of the components and the interfacing of components for structural interaction must be dealt with in design and in the development of proper construction details.

The blending of components of precast and poured-in-place concrete offers some opportunities for complementary enhance-

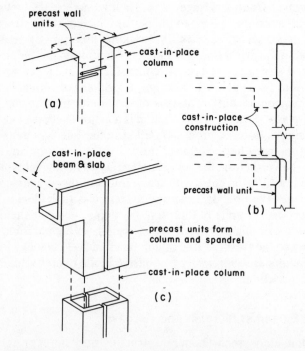

FIGURE 3.3 Mixing of precast and cast-in-place concrete.

ment of the two methods. Figure 3.3*a* shows a joint detail commonly used with tilt-up wall panels. In this case a formed and site-poured concrete column is used to effect the structural connection between two panels, as well as to achieve a positive interaction of the panels (as shear walls) with the continuous poured concrete frame. A similar situation is shown in Fig. 3.3*b*, in which two levels of a structure, consisting of slab-and-beam concrete systems, are connected to a precast concrete wall panel.

A slightly different situation is shown in Fig. 3.3*c*. In this case precast concrete units are used as forms for sitecast concrete columns and the edges of a sitecast framing system. In this case the precast units do not serve significant structural tasks in their own right, in comparison to the situations in Fig. 3.3*a* and *b*. They do, however, eliminate the need for other, temporary forming. More important probably, they provide the possibility for a quality of detail and textural control of the exposed concrete that cannot easily be achieved with sitecast concrete.

3.11 CONCRETE MASONRY

In times past, structural masonry was achieved primarily with stone and bricks. Concrete, in a crude form, was used as filler to allow large elements to be constructed with shells of finer material. At present, in the United States, much of what has the appearance of a masonry structure is really achieved with a thin veneer over some other construction—frequently a wood or steel frame. Structural masonry, when it actually occurs, is sometimes of brick, but most often uses masonry units of precast concrete.

A common form of structural masonry is that shown in Fig. 3.4, called structural reinforced masonry with CMUs (concrete masonry units—good old ''concrete blocks'' to the uninitiated). In this form of construction, concrete is used in two ways: first, for the units that are typically laid with mortar in the time-honored fashion; then, selected vertically aligned voids and horizontal courses of blocks are filled with concrete after steel rods are placed, resulting in the construction of a reinforced concrete rigid frame inside a hollow masonry shell. This form of construction is

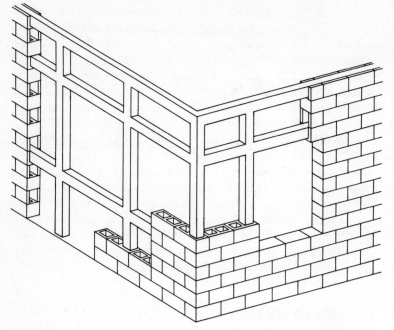

FIGURE 3.4   Use of concrete masonry units with reinforced construction.

most popular in the western and southern areas of the United States.

Structural masonry is used most often for walls, and a discussion of wall design is given in Sec. 10.7. Use of concrete masonry walls is also illustrated in the building system design examples in Chapter 12.

4

GENERAL REQUIREMENTS FOR REINFORCED CONCRETE STRUCTURES

This chapter presents some considerations that generally apply to all concrete structures. Detailed requirements for various types of concrete elements are discussed in the chapters that deal with the general design of those elements. Some considerations for the general development of building structural systems are discussed in Chapters 11 and 12.

4.1 CODE AND INDUSTRY STANDARDS

Standards of practice for structural design and general construction evolve slowly from shared experiences and research. New theories and innovative designs or construction techniques are tested and eventually accepted or rejected. Materials and processes for production and construction are steadily improved. At any point in time, the distilled residue and essence of this is reflected in the latest "standards of practice," presented as building codes and industry standards. Some of the major sources for these standards are discussed in Sec. 1.4.

General practices in design and construction work must acknowledge these basic reference sources. However, published standards are often minimal in nature, specifying the least that must be done, not the best that can be done. Where something more than minimal results are to be expected, satisfying minimal requirements is not likely to be the ideal objective.

Operative standards of practice influence various activities with regard to the design and construction of concrete structures. Major areas of consideration are the following:

Design methods and criteria.

Production and construction processes.

Required tests and certifications.

Code requirements for fire resistance.

General code requirements that affect planning and detailing of building construction.

As mentioned previously, these matters are discussed in detail in many other parts of this book.

4.2 PRACTICAL CONSIDERATIONS

Producing concrete structures is a real-world situation with many very pragmatic concerns. Designers, generally not directly involved in the production work, must relate practically to these concerns or suffer many embarrassments when their speculations face reality. The following are some basic, practical concerns for concrete construction.

Maximum Single Pour

Various practical limits ordinarily establish the maximum amount of concrete that can be placed at one time. For large structures, this is usually only a fraction of the entire structure. Thus the so-called cold joint, or construction joint, occurs when pouring stops for some significant time. The concrete poured hardens, and when pouring continues, the new, wet concrete is poured against the old, hardened concrete. These joints must be anticipated and preferably incorporated into design and detailing of the construction.

The size of pour may be limited by time (the 8-hour workday), by the size of the work crew, by accessibility of the site, by the number of concrete trucks available for delivery, by the method for placing concrete, or by practical limits of the form of construction. An example of the latter is a multistory structure, where a practical pour limit is one story at a time.

For sitecast structures, the entire structure is typically considered as a continuous, monolithic mass. Achieving this in fact requires some careful considerations for the effects of the construction joints.

Concrete Design Strength (f'_c)

For structural design, a major early design decision is that for the design strength of the concrete. This must be related fundamentally to the nature of the structure, but also to practical considerations for the capabilities of the current technology, the abilities of the available workers and contractors, and the size and budget of the project. Thus some designs may push against the limits of the current technology to achieve the best possible concrete (for major high-rise structures), while others may best use a threshold, minimum level of material. This is a very critical issue, but quite difficult to generalize simply.

Accuracy of Construction

Sitecast concrete is very rough work, not generally subject to precise, neat, fine detail and finish. Experience with actual construction will bring some judgment about what must be tolerated and what can be improved with careful specifications, choices for forming materials, and some extra efforts at site inspection of the work.

In general, off-site, plant-cast concrete is capable of achieving higher quality in terms of the concrete material itself, as well as dimensional accuracies and finishes of the cast elements. This may not be so critical to the production of basic structural components, but allows for production of truly superior elements for cladding and other architectural finishing of the construction.

Of course, if the concrete work is to be covered with applied finishes or generally encased in the finished construction, its natural rough character may be of little consequence. Consideration should be given, however, to the tolerances required for attachments of more refined elements of the construction—recognizing the limits of accuracy of the concrete structure.

Minimum Size of Concrete Members

For practical reasons, as well as the satisfying of various requirements for cover and bar spacing, there are minimum usable dimensions for various reinforced concrete members. When flexural reinforcement is required in slabs, walls, or beams, its effectiveness will be determined in part by the distance between the tension-carrying steel and the far edge of the compression carrying concrete. Thus extremely shallow beams and thin slabs or walls will have reduced efficiency for flexure.

In slabs and walls, it is usually necessary to provide two-way reinforcing. Even where the bending actions occur in only one direction, the code requires a minimum amount of reinforcing in the other direction for control of cracking due to shrinkage and temperature changes. As shown in Fig. 4.1a, even with minimum cover and small bars, a minimum slab thickness is approximately 2 in. Except for joist or waffle construction, however, slab thicknesses are usually greater, for reasons of development of practical levels of flexural resistance. Thus reinforcing is more often as shown in Fig. 4.1b, with the bars closer to the top or bottom, depending on whether the moment is positive or negative.

In many cases it is desirable for the slab to have a significant fire rating. Building codes often require additional cover for this purpose, and typically specify minimum slab thicknesses of 4 in. or more. Slab thicknesses required for this purpose also depend on the type of aggregate that is used for the concrete.

Walls of 10 in. or greater thickness often have two separate layers of reinforcing, as shown in Fig. 4.1c. Each layer is placed as close as the requirements for cover permit to the outside wall

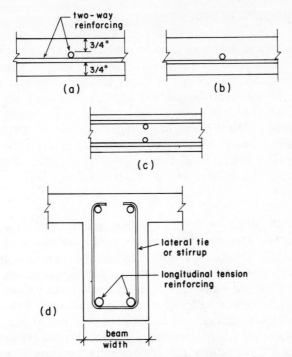

FIGURE 4.1 Placement of steel reinforcement in concrete.

surface. Walls with crisscrossed reinforcing (both vertical and horizontal bars) are seldom made less than 6 in. thick.

As shown in Fig. 4.1d, concrete beams usually have a minimum of two reinforcing bars and a stirrup or tie of at least No. 2 or No. 3 size. Even with small bars, the minimum beam width in this situation is at least 8 in., with 10 in. being much more practical.

For rectangular columns with ties, a limit of 8 in. is usual for one side of an oblong cross section and 10 in. for a square section. Round columns may be either tied or spiral wrapped. A 10-in. diameter may be possible for a round tied column, but 12 in. is more practical and is the usual minimum for a spiral column, with larger sizes required where more cover is necessary.

4.3   CRACK CONTROL

Cracking of concrete structures is a major concern, and the general development of cracks is discussed in Sec. 2.11. Control of cracking is seldom achieved by any single means, but typically by some combination of the following.

Minimum Reinforcement.   Cracks are essentially tension stress failures, and a basic crack-resisting technique is the use of steel reinforcement. Use of minimum reinforcement for resistance to the effects of shrinkage and thermal change is discussed in Sec. 4.4.

Control Joints.   A major means for crack control is the establishment of joints that interrupt the continuity of the concrete mass, in effect functioning as preestablished cracks. Specific considerations for these are discussed in Sec. 9.11 (for paving slabs) and Sec. 10.1 (for walls). Locations of control joints should preferably be developed with logical consideration for practical construction joints (see discussion in Sec. 4.2). Structural design may also indicate the need for control joints for thermal expansion of the general construction, seismic separation, control of continuity effects, or other structural purposes; these joints will also function for crack control. For good coordinated design, any visible joints should be expressed in the general architectural design and detailing of the building.

Fiber Reinforcement.   Although currently used mostly for nonstructural elements of the construction (shingles, cladding panels, etc.), the inclusion of fiber materials in the concrete mix enhances the general tensile resistance of the basic concrete material, resulting in some general reduction of minor cracking.

Prestressing.   As discussed in Sec. 2.8, the use of prestressing is a means for elimination of cracking caused by flexure in beams and other structural components. This is seldom the principal reason for using prestressing, but is nevertheless an added positive factor.

4.4   GENERAL REQUIREMENTS FOR REINFORCEMENT

The following are some general requirements for steel reinforcement in reinforced concrete structures. Specific requirements for
individual types of structural components are also presented in
other parts of this book.

Minimum Reinforcement.   In the design of most reinforced
concrete members the amount of steel reinforcing required is
determined from computations and represents the amount determined to be necessary to resist the required tensile force in the
member. In various situations, however, there is a minimum
amount of reinforcing that is desirable, which may on occasion
exceed that determined by the computations. The ACI Code
makes provisions for such minimum reinforcing in columns,
beams, slabs, and walls. The minimum reinforcing may be specified as a minimum percentage of the member cross-sectional area,
as a minimum number of bars, or as a minimum bar size. These
requirements are discussed in the sections that deal with the design of the various types of members.

Shrinkage and Temperature Reinforcement.   The essential purpose of steel reinforcing is to prevent the cracking of the
concrete due to tension stresses. In the design of concrete structures, investigation is made for the anticipated structural actions
that will produce tensile stress: primarily the actions of bending,
shear, and torsion. However, tension can also be induced by the
shrinkage of the concrete during its drying out after the initial
pour. Temperature variations may also induce tension in various
situations. To provide for these later actions, the ACI Code requires a minimum amount of reinforcing in members such as
walls and slabs even when structural actions do not indicate any
need. These requirements are discussed in the sections that deal
with the design of these members.

Cover.   Cover of the steel bars must be provided for various
reasons. A prime concern is simply the need for the concrete to
"grab" the steel by surrounding it, so that the two materials can

truly interact in the response of the composite structure to needed structural actions. Beyond this are the practical concerns for weather protection, general protection from air and moisture that causes rusting, and insulation for fire protection of the steel. Some code limits for cover are discussed in Sec. 2.7. Specific concerns for particular types of structural components are discussed in other chapters and some detailed concerns are presented in the design examples in Chapter 12.

The outer dimensions of the concrete in a given structural member combine with the required cover dimensions to define a limiting space for the placing of the reinforcement. Within this space, spacing requirements—as discussed next—determine how much reinforcement can actually be placed inside the defined concrete member. If this space is not adequate for the amount of steel required, redesign of the dimensions of the concrete member must usually be considered.

Spacing of Steel Bars. Where multiple bars are used in members (which is the common situation), there are both upper and lower limits for the spacing of the bars. Lower limits are intended to permit adequate development of the concrete-to-steel stress transfers and to facilitate the flow of the wet concrete during pouring. For columns, the minimum clear distance between bars is specified as 1.5 times the bar diameter or a minimum of 1.5 in. For other situations, the minimum is one bar diameter or a minimum of 1 in.

For walls and slabs, maximum center-to-center bar spacing is specified as three times the wall or slab thickness or a maximum of 18 in. This applies to reinforcement required for computed stresses. For reinforcement that is required for control of cracking due to shrinkage or temperature change, the maximum spacing is five times the wall or slab thickness or a maximum of 18 in.

For adequate placement of the concrete, the largest size of the coarse aggregate should be not greater than three-fourths of the clear distance between bars.

Bending of Reinforcement. In various situations, it is sometimes necessary to bend reinforcing bars. Bending is done preferably in the fabricating shop instead of at the job site, and the

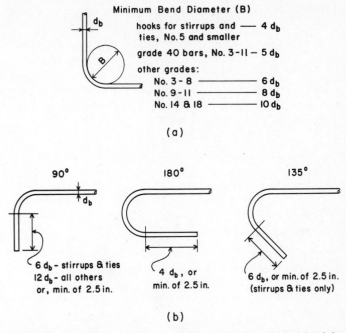

(a)

(b)

FIGURE 4.2 Bend requirements for steel reinforcing bars: (a) minimum bend diameters; (b) requirements for standard hooks.

bend diameter (see Fig. 4.2) should be adequate to avoid cracking the bar.

Bending of bars is sometimes done in order to provide anchorage for the bars. The code defines such a bend as a "standard hook," and the requirements for the details of this type of bend are given in Fig. 4.2b.

As the yield stress of the steel is raised, bending becomes increasingly difficult. Bending of bars should be avoided when the yield stress exceeds 60 ksi [414 MPa]; and where it is necessary, should be done with bend diameters slightly greater than those given in Fig. 4.2.

5

INVESTIGATION AND DESIGN
OF REINFORCED CONCRETE

This chapter presents some of the basic considerations in the design of reinforced concrete structures and the necessary investigations of structural behaviors to support the design work.

5.1 GENERAL CONCERNS FOR CONCRETE

In reinforced concrete the concrete itself is relied on primarily only for resistance of compressive stress. Its limitation in this regard is defined by the assumed design strength (f'_c), which is established essentially by a compression test on the material. Almost all other structural properties are based on this defined strength limit. The major resistance to tension is assigned to the steel reinforcement, so that investigations of the concrete are limited essentially to concerns for maximum compressive stress conditions.

Because of the nature of interaction of the two materials (concrete and steel) in the composite reinforced concrete structure, stress distributions between the materials are affected by their relative stiffness, as indicated by the modulus of elasticity of the materials. The modulus of the steel remains constant through all grades of the reinforcement. However, the modulus of the concrete changes, as discussed in Sec. 2.6. For this purpose, as well as any investigations of structural deformations, the concrete modulus of elasticity must also be established.

5.2 GENERAL CONCERNS FOR REINFORCEMENT

Most steel used for reinforcement is highly ductile in nature. Its usable strength is its yield strength, as this stress condition initiates such a magnitude of deformation (into the plastic yielding range of the steel), that major cracking will occur in the concrete. Since the yield strength of the steel is quite clearly defined and controlled, this establishes a very precise reference in structural investigations. Reference to this is made in later sections in this chapter. An early design decision is that for the yield strength (specified by the grade of steel used) that is to be used in the design work.

Several different grades of steel may be used for large projects, with a minimum grade for ordinary tasks and higher grades for more demanding ones. Cost increases generally for higher grades, so some feasibility studies must be made to see if the better steel in smaller quantities is really cheaper than a larger quantity of a lower grade. Actually, higher grades are often used to permit smaller concrete members, relating to the space problems for placement of the reinforcement, as discussed in Sec. 4.4.

Even though the steel ordinarily constitutes only a few percent of the total volume of reinforced concrete, it is a major cost factor. This includes the cost of the steel, the forming of the deformed bars, the cutting and bending required, and the installation in the forms. A cost-saving factor is usually represented by the general attempt to use the minimum reinforcement and the most concrete, reflecting typical unit costs for the two materials.

5.3 GENERAL CONCERNS FOR CONCRETE STRUCTURES

In most situations it is desired that the structural failure of a concrete member be initiated by yielding of the ductile reinforcement. There are a number of reasons for this, one being the much more reliable and predictable value for strength of the steel versus that of the concrete. This character for the reinforced concrete is imaged by striving to have members generally under-reinforced rather than over-reinforced. Thus the failure of the steel occurs before the concrete approaches its stress limits.

Many major structural members, however, are reinforced with both tensile and compressive reinforcement; thus it is not simply a case of the compressive concrete versus the tensile steel. Nevertheless, the use of the yielding of the steel as the indicator of strength limits is still the desired condition. Another reason for preferring this mode of failure is that the failure of the concrete— in either compressive crushing or tensile cracking—is generally in the nature of a sudden failure. The yielding steel (deforming plastically), on the other hand, retains some strength—giving it an aspect of toughness. The tough structure is much to be preferred over the brittle one.

The general idea in design of reinforced concrete is to assure that concrete members are adequate in size (outer dimensions), and that steel is placed in the proper places, in the necessary amount, and with the linear bars oriented to resist the tension required.

5.4 INVESTIGATION METHODS

As discussed in Sec. 1.5, there are two fundamentally different methods for investigation and design of reinforced concrete: the working stress method and the strength method. The strength method is now used almost exclusively for professional design work and is the basis for development of the various computer-assisted processes now in wide use.

For study purposes, both methods are presented in this book.

The design examples mostly use the working stress method simply because its procedures and formulizations are somewhat simpler and less abstract for visualization. Many aspects of design of structures do not relate to loading or stress investigations, so that they apply equally to processes that use either method of investigation. Those readers who intend to go on to serious professional design work should pursue more rigorous study of structural investigation in general, including the use of the strength methods.

5.5 THE WORKING STRESS METHOD

As applied to the investigation of behaviors and the design of members of reinforced concrete, the working stress method consists of the determination of stresses in members that are induced by the actual loading under working conditions—called service load conditions. The stresses thus determined are then compared to the limits established for the situation under investigation. These limiting stresses—called allowable stresses—are established by code requirements as are the methods by which the actual stresses are determined. If the actual stresses do not exceed the allowable stresses, the member is considered to be adequate.

For concrete, allowable stresses are essentially based on the established design strength of the material. This strength is the so-called ultimate compressive strength, designated as f'_c, which is determined from the testing of standard samples, as discussed in Chapter 3.

Allowable stresses for steel are based on the yield strength of the steel, designated as f_y. Formulas used for determination of actual stresses in both the concrete and the steel are essentially derived from consideration of elastic behavior of members. In many cases, however, adjustments are made on the purely elastic formulas to account for the nonlinear stress-strain behavior of the concrete.

The working stress method is no longer favored by the codes and has largely been replaced in professional design practice by strength design methods. It is not our purpose to advocate the use of one method over the other; both methods are presented for most of the work in this book. In general the working stress method is simpler to explain and its methods are easier and less complex to use. In some situations it is still used by designers and is allowed by most building codes. The latest edition of the ACI Code (1989 ed.) still provides for its use in a limited number of situations.

As applied to the work in this book, the working stress method is based primarily on the current requirements of the ACI Code. These are described in Appendix A of the code, under the title Alternate Method. The last edition of the code to develop the method in a full manner was the 1963 edition, and some of the material used here is taken from that publication.

5.6 THE STRENGTH METHOD

Application of the working stress method consists of designing members to *work* in an adequate manner (without exceeding established stress limits) under actual service load conditions. The basic procedure in strength design is to design members to *fail*; thus the ultimate strength of the member at failure (called its design strength) is the only type of resistance considered. Safety in strength design is not provided by limiting stresses, as in the working stress method, but by using a factored design load (called the *required strength*) that is greater than the service load. The code establishes the value of the required strength, called U, as not less than

$$U = 1.4D + 1.7L \tag{1}$$

where D = effect of dead load

L = effect of live load

Other adjustment factors are provided when design conditions involve consideration of the effects of wind, earth pressure, differential settlement, creep, shrinkage, or temperature change.

The design strength of structural members (i.e., their *usable* ultimate strength) is determined by the application of assumptions and requirements given in the code and is further modified by the use of a *strength reduction factor* ϕ as follows:

ϕ = 0.90 for flexure, axial tension, and combinations of flexure and tension

= 0.75 for columns with spirals

= 0.70 for columns with ties

= 0.85 for shear and torsion

= 0.70 for compressive bearing

Thus while formula (1) may imply a relatively low safety factor, an additional margin of safety is provided by the stress reduction factors.

5.7 INVESTIGATION OF BEAMS AND FRAMES

Structural investigation begins with an overall analysis of the entire structure to determine responses at supports (reactions) and the types and magnitudes of interior force actions. For systems that are comprised of simple beams and single, pin-ended columns (as in most wood frames, for example), this analysis is quite easily accomplished. Most concrete structures, on the other hand, have members that are continuous through many spans and beams and columns that constitute rigid frames; they are thus very statically indeterminate and their analysis for resistive forces is quite complex. Analysis of indeterminate structures in general is beyond the scope of this book, but the following discussions are provided to explain some elementary ideas and procedures.

Investigation of Beams

The simple, single-span beam is a rare situation in reinforced concrete structures. As shown in Fig. 5.1, the simple beam may exist when a single span is supported on bearing-type supports that offer little restraint (Fig. 5.1*a*), or when beams are connected

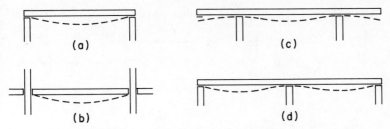

FIGURE 5.1 Flexural action of various beam forms.

to columns with connections that offer little moment resistance (Fig. 5.1b). Although these situations are common in structures of steel and wood, they seldom occur in concrete structures, except when precast elements are used.

For single-story structures, supported on bearing-type supports, continuity resulting in complex bending can occur when the spanning members are extended over the supports. This may occur in the form of cantilevered ends (Fig. 5.1c), or of multiple spans (Fig. 5.1d). These conditions are common in wood and steel structures, and can also occur in reinforced concrete structures. Members of steel and wood are usually constant in cross section throughout their length; thus it is necessary only to find the single maximum value for shear and the single maximum value for moment. For the concrete member, however, the variations of shear and moment along the beam length must be considered, and several different cross sections must be investigated.

Figure 5.2 shows conditions that are common in concrete structures when beams and columns are cast monolithically. For the single-story structure (Fig. 5.2a), the rigid joint between the beam and its supporting columns will result in behavior shown in Fig. 5.2b; with the columns offering some degree of restraint to the rotation of the beam ends. Thus some moment will be added to the tops of the columns and the beam will behave as for the center portion of the span in Fig. 5.1c, with both positive and negative moments.

For the multiple-story, multiple-span concrete frame, the typical behavior will be as shown in Figs. 5.2c and 5.2d. The columns above and below, plus the beams in adjacent spans, will contrib-

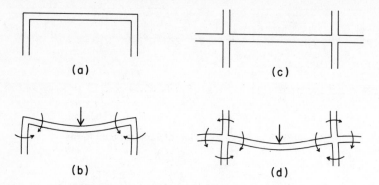

FIGURE 5.2 Actions of beams in rigid frames.

ute to the development of restraint for the ends of an individual beam span. This condition occurs in steel structures only when welded or heavily bolted moment-resisting connections are used. In concrete structures, it is the normal condition.

The structures shown in Figs. 5.1d, 5.2a, and 5.2c are statically indeterminate. This means that their investigation cannot be performed using only the conditions of static equilibrium. Although a complete consideration of statically indeterminant behaviors is well beyond the scope of this book, some treatment must be given for a realistic development of the topic of design of reinforced concrete structures. The discussions that follow will serve to illustrate the various factors in the behavior of continuous frames and will provide material for approximate analysis of common situations.

Effects of Beam End Restraint

Figure 5.3a to d shows the effects of various end support conditions on a single-span beam with a uniformly distributed load. Similarly, Figure 5.3e to h shows the conditions for a beam with a single concentrated load. Values are indicated for the maximum shears, moments, and deflection for each case. (Values for end reaction forces are not indicated, since they are the same as the end shears.)

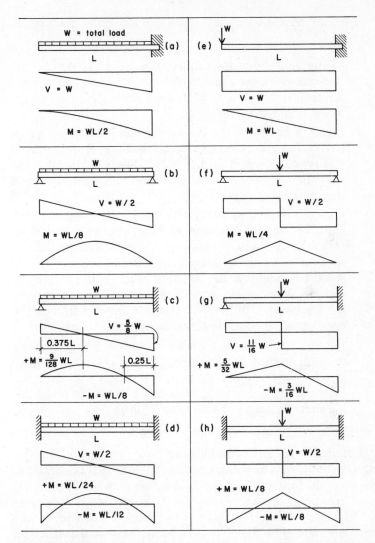

FIGURE 5.3 Beam response values for beams with uniformly distributed loading and single concentrated loading.

We note the following for the four cases of end support conditions.

1. Figure 5.3a shows the cantilever beam, supported at only one end with a *fixed-end* condition. Both shear and moment are critical at the fixed end, and maximum deflection occurs at the unsupported end.

2. Figure 5.3b shows the classic "simple" beam, with supports offering only vertical force resistance. We will refer to this type of support as a *free end*. Shear is critical at the supports and both moment and deflection are maximum at the center of the span.

3. Figure 5.3c shows a beam with one free end and one fixed end. This support condition produces an unsymmetrical situation for the vertical reactions and the shear. The critical shear occurs at the fixed end, but both ends must be investigated separately for the concrete beam. Both positive and negative moments occur, with the maximum moment being the negative one at the fixed end. Maximum deflection will occur at some point slightly closer to the free end.

4. Figure 5.3d shows the beam with both ends fixed. This symmetrical support condition results in a symmetrical situation for the reactions, shear, and moments with the maximum deflection occurring at midspan. It may be noted that the shear diagram is the same as for the simple beam in Fig. 5.3b.

Continuity and end restraint have both positive and negative effects with regard to various considerations. The most positive gain is in the form of reduction of deflections, which is generally more significant for steel and wood structures, since deflections are less often critical for concrete members. For the beam with one fixed end (Fig. 5.3c), it may be noted that the value for maximum shear is increased and the maximum moment is the same as for the simple span (no gain in those regards). For full end fixity (Fig. 5.3d), the shear is unchanged, while both moment and deflection are quite substantially reduced in magnitude.

For the rigid frames shown in Fig. 5.2, the restraints will reduce moment and deflection for the beam, but the cost is at the expense of the columns, which must take some moment in addition to axial force. Rigid frames are often utilized to resist lateral loads due to wind and earthquakes, presenting complex combinations of lateral and gravity loading that must be investigated.

Effects of Concentrated Loads

Framing systems for roofs and floors often consist of series of evenly spaced beams that are supported by other beams placed at right angles to them. The supporting beams are thus subjected to a series of spaced, concentrated loads—the end reactions of the supported beams. The effects of a single such load at the center of a beam span are shown in Fig. 5.3*e* to *h*. Two additional situations of evenly spaced concentrated loading are shown in Fig. 5.4. When more than three such loads occur, it is usually adequate to consider the sum of the concentrated loads as a uniformly distributed load and to use the values given for Fig. 5.3*b* to *d*.

Multiple Beam Spans

Figure 5.5 shows various loading conditions for a beam that is continuous through two equal spans. When continuous spans occur, it is usually necessary to give some consideration to the possibilities of partial beam loading, as shown in Fig. 5.5*b* and *d*. It may be noted for Fig. 5.5*b* that although there is less total load on the beam, the values for maximum positive moment, deflection, and shear at the free end are all higher than for the fully loaded beam in Fig. 5.5*a*. This condition of partial loading must be considered for *live loads* (people, furniture, snow, etc.). For design, the partial loading effects due to the live load must be combined with those produced by dead load (permanent weight of the construction) for the full action of the beam.

Figure 5.6 shows a beam that is continuous through three equal spans, with various situations of uniform load on the beam spans. Figure 5.6*a* gives the loading condition for dead load (*always* present in *all* spans). Figure 5.6*b* to *d* show the several possibilities for partial loading, each of which produces some specific critical values for the reactions, shears, moments, and deflections.

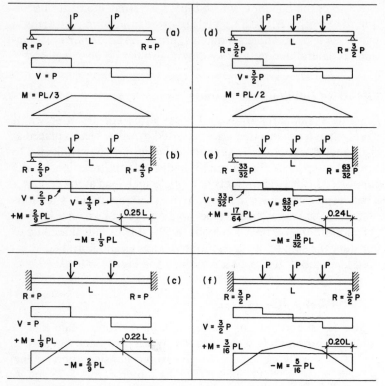

FIGURE 5.4   Beam response values for multiple concentrated loads.

Complex Loading and Span Conditions

Although values have been given for many common situations in Figs. 5.3 to 5.6, there are numerous other possibilities in terms of unsymmetrical loadings, unequal spans, cantilevered free ends, and so on. Where these occur, an analysis of the indeterminate structure must be performed. For some additional conditions, the reader is referred to various handbooks that contain tabulations similar to those presented here. Two such references are the *CRSI Handbook* (Ref. 3) and the AISC Manual (*Manual of Steel Construction*, 8th ed., published by the American Institute of Steel Construction).

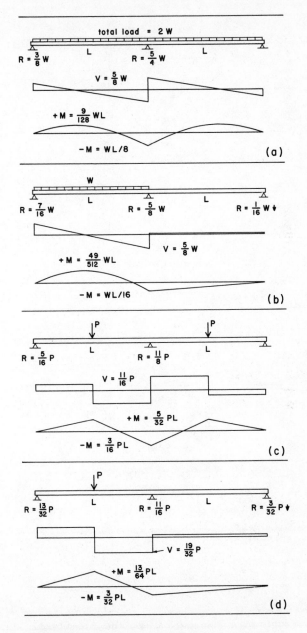

FIGURE 5.5   Response values for two-span beams.

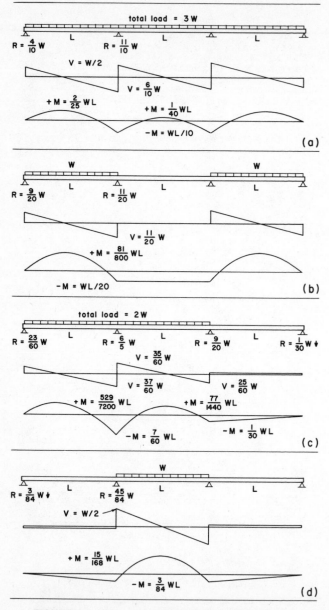

FIGURE 5.6 Response values for three-span beams.

5.8 RIGID FRAMES

Frames in which two or more of the members are attached to each other with connections that are capable of transmitting bending between the ends of the members are called *rigid frames*. The connections used to achieve such a frame are called *moment connections* or *moment-resisting connections*. Most rigid frame structures are statically indeterminate and do not yield to investigation by consideration of static equilibrium alone. The examples presented in this section are all rigid frames that have conditions that make them statically determinate and thus capable of being fully investigated by methods developed in this book.

Cantilever Frames

Consider the frame shown in Fig. 5.7a, consisting of two members rigidly joined at their intersection. The vertical member is fixed at its base, providing the necessary support condition for stability of the frame. The horizontal member is loaded with a uniformly distributed loading and functions as a simple cantilever beam. The frame is described as a cantilever frame because of the single fixed support. The five sets of figures shown in Fig. 5.7b through f are useful elements for the investigation of the behavior of the frame:

1. The free-body diagram of the entire frame, showing the loads and the components of the reactions (Fig. 5.7b). Study of this figure will help in establishing the nature of the reactions and in the determination of the conditions necessary for stability of the frame.
2. The free-body diagrams of the individual elements (Fig. 5.7c). These are of great value in visualizing the interaction of the parts of the frame. They are also useful in the computations for the internal forces in the frame.
3. The shear diagrams of the individual elements (Fig. 5.7d). These are sometimes useful for visualizing, or for actually computing, the variations of moment in the individual elements. No particular sign convention is necessary unless in

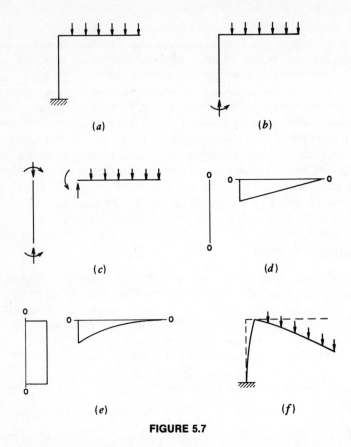

FIGURE 5.7

conformity with the sign used for moment. Although good as exercises in visualization, the shear diagrams have limited value in the investigation in most cases.

4. The moment diagrams for the individual elements (Fig. 5.7e). These are very useful, especially in determination of the deformation of the frame. The sign convention used is that of plotting the moment on the compression side of the element.

5. The deformed shape of the loaded frame (Fig. 5.7f). This is the exaggerated profile of the bent frame, usually superimposed on an outline of the unloaded frame for reference.

This is very useful for the general visualization of the frame behavior. It is particularly useful for determination of the character of the external reactions and the form of interaction between the parts of the frame.

When performing investigations, these elements are not usually produced in the sequence just described. In fact, it is generally recommended that the deformed shape be sketched first so that its correlation with other factors in the investigation may be used as a check on the work. The following examples illustrate the process of investigation for simple cantilever frames.

Example 1. Find the components of the reactions and draw the free-body diagrams, shear and moment diagrams, and the deformed shape of the frame shown in Fig. 5.8*a*.

Solution: The first step is the determination of the reactions. Considering the free-body diagram of the whole frame (Fig. 5.8*b*), we compute the reactions as follows:

$$\Sigma F = 0 = +8 - R_V, \qquad R_V = 8 \text{ kips (up)}$$

and with respect to the support labeled 0,

$$\Sigma M_0 = 0 = M_R - (8 \times 4), \quad M_R = 32 \text{ kip-ft (counterclockwise)}$$

Note that the sense, or sign, of the reaction components is visualized from the logical development of the free-body diagram.

Consideration of the free-body diagrams of the individual members will yield the actions required to be transmitted by the moment connection. These may be computed by application of the conditions for equilibrium for either of the members of the frame. Note that the sense of the force and moment is opposite for the two members, simply indicating that what one does to the other is the opposite of what is done to it.

In this example there is no shear in the vertical member. As a result, there is no variation in the moment from the top to the bottom of the member. The free-body diagram of the member, the shear and moment diagrams, and the deformed shape should all

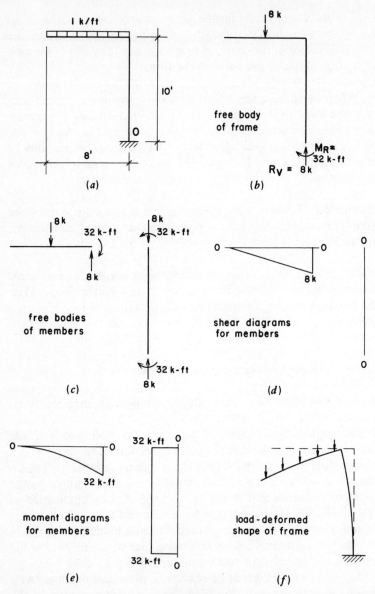

FIGURE 5.8

corroborate this fact. The shear and moment diagrams for the horizontal member are simply those for a cantilever beam.

It is possible with this example, as with many simple frames, to visualize the nature of the deformed shape without recourse to any mathematical computations. It is advisable to do so, and to check continually during the work that individual computations are logical with regard to the nature of the deformed structure.

Example 2. Find the components of the reactions and draw the shear and moment diagrams and the deformed shape of the frame in Fig. 5.9*a*.

Solution: In this frame there are three reaction components required for stability, since the loads and reactions constitute a general coplanar force system. Using the free-body diagram of the whole frame (Fig. 5.9*b*), the three conditions for equilibrium for a coplanar system are used to find the horizontal and vertical reaction components and the moment component. If necessary, the reaction force components could be combined into a single-force vector, although this is seldom required for design purposes.

Note that the inflection occurs in the larger vertical member because the moment of the horizontal load about the support is greater than that of the vertical load. In this case, this computation must be done before the deformed shape can be accurately drawn.

The reader should verify that the free-body diagrams of the individual members are truly in equilibrium and that there is the required correlation between all the diagrams.

Single-Bent Frames

Figure 5.10 shows two possibilities for a single-bent rigid frame. In Fig. 5.10*a* the frame has pinned bases for the columns, resulting in the form of deformation under loading as shown in Fig. 5.10*c*, and the reaction components as shown in the free-body diagram for the whole frame in Fig. 5.10*e*. The second frame (Fig. 5.10*b*) has fixed bases for the columns, resulting in the slightly modified behavior indicated. These are the two most common

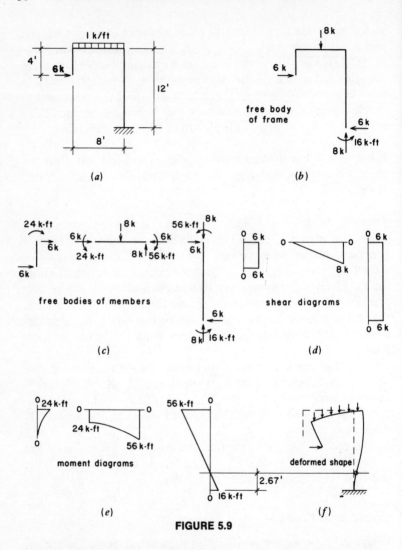

FIGURE 5.9

forms of single-bent frames, the choice of the column base condition depending on a number of design factors for each case.

Both of the frames shown in Fig. 5.10 are statically indeterminate, and their investigation is beyond the scope of work in this book. The following examples of single-bent frames consist of frames with combinations of support and internal conditions that

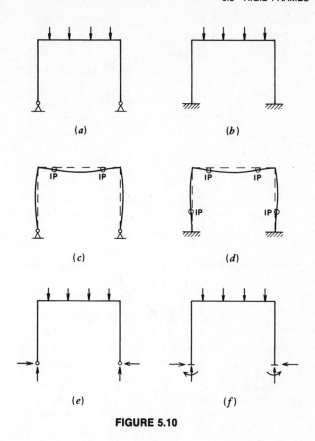

FIGURE 5.10

make the frames statically determinate. These conditions are technically achievable, but a bit on the weird side for practical use. We offer them here simply as exercises within the scope of our readers so that some experience in investigation may be gained.

Example 3. Investigate the frame shown in Fig. 5.11 for the reactions and internal conditions. Note that the right-hand support allows for an upward vertical reaction only, whereas the left-hand support allows for both vertical and horizontal components. Neither support provides moment resistance.

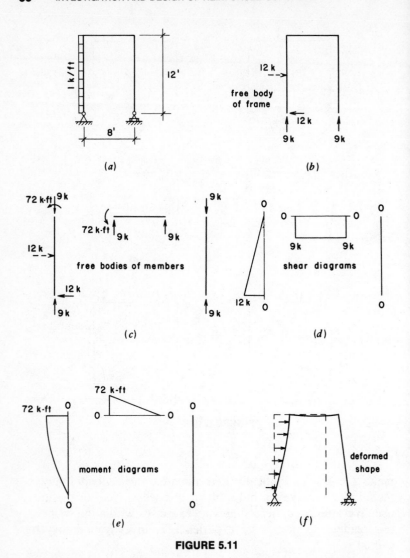

FIGURE 5.11

Solution: The typical elements of investigation, as illustrated for the preceding examples, are shown in the figure. The suggested procedure for the work is as follows:

1. Sketch the deflected shape (a little tricky in this case, but a good exercise).

2. Consider the equilibrium of the free-body diagram for the whole frame to find the reactions.
3. Consider the equilibrium of the left-hand vertical member to find the internal actions at its top.
4. Proceed to the equilibrium of the horizontal member.
5. Finally, consider the equilibrium of the right-hand vertical member.
6. Draw the shear and moment diagrams and check for correlation of all the work.

Before attempting the exercise problems, the reader is advised to attempt to produce the results shown in Fig. 5.11 independently.

PROBLEMS 5.8.A,B,C. For the frames shown in Fig. 5.12*a* through *c*, find the components of the reactions, draw the free-body diagrams of the whole frame and the individual members, draw the shear and moment diagrams for the individual members, and sketch the deformed shape of the loaded structure.

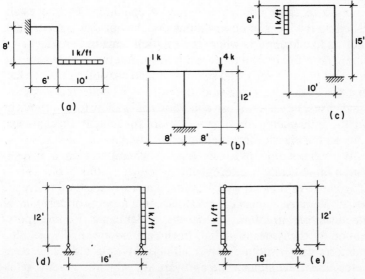

FIGURE 5.12

PROBLEMS 5.8.D,E. Investigate the frames shown in Fig. 5.12*d* and *e* for reactions and internal conditions, using the procedure shown for the preceding examples.

5.9 APPROXIMATE ANALYSIS OF INDETERMINATE FRAMES

Analysis of the behavior of indeterminate structures requires use of some conditions in addition to those provided by consideration of static equilibrium. They are thus imaged as having a negative character: being *not* capable of something—analysis by statics alone. Actually, there is nothing wrong or bad about a structure that is statically indeterminate. In fact, there are several potential advantages, including reduction of the maximum bending moments, reduction of deformations, and a general redundancy of stability and internal resisting forces that enhances safety. In some cases the structure is deliberately made redundant to gain one or more of these advantages. In other cases the structure is inherently indeterminate due to the nature of the construction— such as with cast-in-place concrete structures with multiple spans.

In some cases additional behavior conditions can be established on the basis of observations of behavior that is controlled, such as fixed supports where the rotation remains zero, any support where the deflection remains zero, internal pins where the moment must be zero, and so on. It is also possible to use the fact that the beam assumes a smooth curve as its deflected shape. In most cases, however, these conditions only allow for shortcuts or for the implementation of other, more fundamental techniques, such as the classic slope-deflection method.

We will not attempt to develop the general problem of investigation of statically indeterminate structures in this book. This is an exhaustive subject, well developed in many standard reference texts. We must, however, consider some aspects of behavior of indeterminate structures so that the reader can gain some appreciation for the issues involved. In current practice really complex indeterminate structures are routinely investigated using computer-aided techniques, the software for which is readily available (for a price) to the design professions.

The rigid-frame structure occurs quite frequently as a multiple-level, multiple-span bent, constituting part of the structure for a multistory building. In most cases such a bent is used as a lateral bracing element, although once it is formed as a moment-resistive framework it will respond characteristically for all types of loads.

The multistory rigid bent is quite indeterminate, and its investigation is complex—requiring considerations of several different loading combinations. When loaded or formed unsymmetrically, it will experience sideways movements that further complicate the analysis for internal forces. Except for very early design approximations, the analysis is now sure to be done with a computer-aided system. The software for such a system is quite readily available.

For preliminary design purposes, it is sometimes possible to use approximate analysis methods to obtain member sizes of reasonable accuracy. Actually, many of the older high-rise buildings still standing were completely designed with these techniques—a reasonable testimonial to their effectiveness. Demonstrations of these approximate methods are given in Chapter 12.

6

REINFORCED CONCRETE
BEAMS

This chapter presents materials relating to the design of concrete beams with conventional reinforcement. General concentration is on beams as they occur in monolithic slab and beam systems, although a general discussion of that form of system and others for achieving flat spans is given in Chapter 7.

6.1 GENERAL BEAM ACTIONS

The primary concerns for beams relate to their necessary resistance to bending and shear and some limitation on their deflection. For wood and steel beams it is usually only necessary to consider the maximum values for shear and moment on a single beam. Concrete beams, on the other hand, are typically multiple-span or members of frames, and thus must be investigated for the entire range of moments and shears along the beam length. This requires considerations for the concrete member dimensions and required reinforcement at many points along the beam. This process will be illustrated more fully in Secs. 7.2 through 7.4.

6.2   BEHAVIOR OF CONCRETE BEAMS

Behaviors of concrete beams have been exhaustively studied in laboratory tests on actual beams. These test data have been compared to analytical theories and used to adjust the investigation and design processes to reflect some conformance with reality. Beam actions are generally visualized in simple terms and many formulas are used in quite simple form, but data used have often been adjusted to assure some conformance with real situations as ordinarily encountered in structures and the quite complex stress actions in the composite concrete + steel elements.

Code design criteria is often quite complex and extensive and assumes that the user fully understands both the theories and realities of reinforced concrete structures. Although the ACI Code is now published in tandem with extensive commentary, it is still not really informative for untrained and inexperienced persons. In this book there is extensive illustration of the investigation and design processes, presented in the interest of aiding the reader to visualize the relationships presented almost entirely in writing and tabulations in the design codes and standards.

Readers who have not studied general structural theory are advised to read some reference to gain background in the topics of beam actions, properties of sections, and the general development of shear and flexure in beams. References 11 and 12 are two books containing such materials, presented for the reader with limited engineering training, as is the material in this book.

6.3   DEVELOPMENT OF FLEXURE

When a member is subjected to bending, such as the beam shown in Fig. 6.1a, internal resistances of two basic kinds are generally required. Internal actions are "seen" by visualizing a cut section, such as that taken at $X-X$ in Fig. 6.1a. Removing the portion of the beam to the left of the cut section, we visualize its free-body actions as shown in Fig. 6.1b. At the cut section, consideration of static equilibrium requires the development of the internal shear force (V in the figure) and the internal resisting moment (represented by the force couple: C and T in the figure).

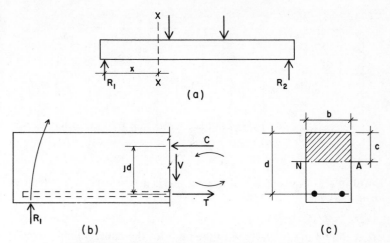

FIGURE 6.1 Bending in a reinforced concrete beam.

If the beam consists of a simple rectangular concrete section with tension reinforcing only, as shown in Fig. 6.1c, the force C is considered to be developed by compressive stresses in the concrete—indicated by the shaded area above the neutral axis. The tension force, however, is considered to be developed by the steel alone, ignoring the tensile resistance of the concrete. For low-stress conditions the latter is not true, but at a serious level of stress the tension-weak concrete will indeed crack, virtually leaving the steel unassisted, as assumed.

At moderate levels of stress, the resisting moment is visualized as shown in Fig. 6.2a, with a linear variation of compressive stress from zero at the neutral axis to a maximum value of f_c at the edge of the section. As stress levels increase, however, the nonlinear stress–strain character of the concrete becomes more significant, and it becomes necessary to acknowledge a more realistic form for the compressive stress variation, such as that shown in Fig. 6.2b. As stress levels approach the limit of the concrete, the compression becomes vested in an almost constant magnitude of unit stress, concentrated near the top of the section. For strength design, in which the moment capacity is expressed as the ultimate limit, it is common to assume the form of stress distribu-

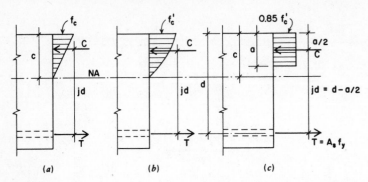

FIGURE 6.2 Development of internal bending resistance in a reinforced concrete beam.

tion shown in Fig. 6.2c, with the limit for the concrete stress set at 0.85 times f'_c. Expressions for the moment capacity derived from this assumed distribution have been shown to compare reasonably with the response of beams tested to failure in laboratory experiments.

Response of the steel reinforcing is more simply visualized and expressed. Since the steel area in tension is concentrated at a small location with respect to the size of the beam, the stress in the bars is considered to be a constant. Thus at any level of stress the total value of the internal tension force may be expressed as

$$T = A_s f_s$$

and for the practical limit of T,

$$T = A_s f_y$$

6.4   INVESTIGATION AND DESIGN: WORKING STRESS METHOD

In working stress design a maximum allowable (working) value for the extreme fiber stress is established (Table 2.1) and the formulas are predicated on elastic behavior of the reinforced concrete member under service load. The straight-line distribution of

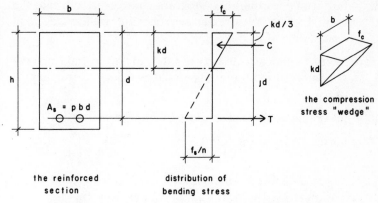

FIGURE 6.3 Bending resistance: working stress method.

compressive stress is valid at working stress levels because the stresses developed vary approximately with the distance from the neutral axis, in accordance with elastic theory.

Flexural Formulas: Working Stress Method

The following is a presentation of the formulas and procedures used in the working stress method. The discussion is limited to a rectangular beam section with tension reinforcing only.

Referring to Fig. 6.3, the following are defined:

b = width of the concrete compression zone

d = effective depth of the section for stress analysis; from the centroid of the steel to the edge of the compression zone

A_s = cross-sectional area of the reinforcing

p = percentage of reinforcing, defined as

$$p = \frac{A_s}{bd}$$

n = elastic ratio = $\dfrac{E \text{ of the steel reinforcing}}{E \text{ of the concrete}}$

kd = height of the compression stress zone; used to locate the neutral axis of the stressed section; expressed as a decimal fraction (k) of d

jd = internal moment arm, between the net tension force and the net compression force; expressed as a decimal fraction (j) of d

f_c = maximum compressive stress in the concrete

f_s = tensile stress in the reinforcing

The compression force C may be expressed as the volume of the compression stress "wedge," as shown in the figure.

$$C = \tfrac{1}{2}(kd)(b)(f_c) = \tfrac{1}{2}kf_cbd$$

Using the compression force, we may express the moment resistance of the section as

$$M = Cjd = (\tfrac{1}{2}kf_cbd)(jd) = \tfrac{1}{2}kjf_cbd^2 \qquad (1)$$

This may be used to derive an expression for the concrete stress:

$$f_c = \frac{2M}{kjbd^2} \qquad (2)$$

The resisting moment may also be expressed in terms of the steel and the steel stress as

$$M = Tjd = A_sf_sjd$$

This may be used for determination of the steel stress or for finding the required area of steel:

$$f_s = \frac{M}{A_sjd} \qquad (3)$$

$$A_s = \frac{M}{f_sjd} \qquad (4)$$

A useful reference is the so-called *balanced section*, which occurs when use of the exact amount of reinforcing results in the simultaneous limiting stresses in the concrete and steel. The properties that establish this relationship may be expressed as follows:

$$\text{balanced } k = \frac{1}{1 + f_s/nf_c} \tag{5}$$

$$j = 1 - \frac{k}{3} \tag{6}$$

$$p = \frac{f_c k}{2f_s} \tag{7}$$

$$M = Rbd^2 \tag{8}$$

in which

$$R = \tfrac{1}{2}kjf_c \tag{9}$$

derived from formula (1). If the limiting compression stress in the concrete ($f_c = 0.45f'_c$) and the limiting stress in the steel are entered in formula (5), the balanced section value for k may be found. Then the corresponding values for j, p, and R may be found. The balanced p may be used to determine the maximum amount of tensile reinforcing that may be used in a section without the addition of compressive reinforcing. If less tensile reinforcing is used, the moment will be limited by the steel stress, the maximum stress in the concrete will be below the limit of $0.45f'_c$, the value of k will be slightly lower than the balanced value, and the value of j will be slightly higher than the balanced value. These relationships are useful in design for the determination of approximate requirements for cross sections.

Table 6.1 gives the balanced section properties for various combinations of concrete strength and limiting steel stress. The values of n, k, j, and p are all without units. However, R must be expressed in particular units; the units used in the table are kip-inch (kip-in.) and kilonewton-meters (kN-m).

When the area of steel used is less than the balanced p, the true value of k may be determined by the following formula:

$$k = \sqrt{2np - (np)^2} - np \tag{10}$$

TABLE 6.1 Balanced Section Properties for Rectangular Concrete Sections with Tension Reinforcing Only

f_s		f'_c						R	
ksi	MPa	ksi	MPa	n	k	j	p	kip-in.	kN-m
16	110	2.0	13.79	11.3	0.389	0.870	0.0109	0.152	1045
		2.5	17.24	10.1	0.415	0.862	0.0146	0.201	1382
		3.0	20.68	9.2	0.437	0.854	0.0184	0.252	1733
		4.0	27.58	8.0	0.474	0.842	0.0266	0.359	2468
20	138	2.0	13.79	11.3	0.337	0.888	0.0076	0.135	928
		2.5	17.24	10.1	0.362	0.879	0.0102	0.179	1231
		3.0	20.68	9.2	0.383	0.872	0.0129	0.226	1554
		4.0	27.58	8.0	0.419	0.860	0.0188	0.324	2228
24	165	2.0	13.79	11.3	0.298	0.901	0.0056	0.121	832
		2.5	17.24	10.1	0.321	0.893	0.0075	0.161	1107
		3.0	20.68	9.2	0.341	0.886	0.0096	0.204	1403
		4.0	27.58	8.0	0.375	0.875	0.0141	0.295	2028

Figure 6.4 may be used to find approximate k values for various combinations of p and n. Beams with reinforcement less than that required for the balanced moment are called *underbalanced sections* or *under-reinforced sections*. If a beam must carry bending moment in excess of the balanced moment for the section, it is necessary to provide some compressive reinforcement, as discussed in Sec. 6.6. The balanced section is not necessarily a design ideal, but is useful in establishing the limits for the section.

In the design of concrete beams, there are two situations that commonly occur. The first occurs when the beam is entirely undetermined; that is, the concrete dimensions and the reinforcing are unknown. The second occurs when the concrete dimensions are given, and the required reinforcing for a specific bending moment must be determined. The following examples illustrate the use of the formulas just developed for each of these problems.

Example 1. A rectangular concrete beam of concrete with f'_c of 3000 psi [20.7 MPa] and steel reinforcing with $f_s = 20$ ksi [138 MPa] must sustain a bending moment of 200 kip-ft [271 kN-m]. Select the beam dimensions and the reinforcing for a section with tension reinforcing only.

Solution: (1) With tension reinforcing only, the minimum size beam will be a balanced section, since a smaller beam would have

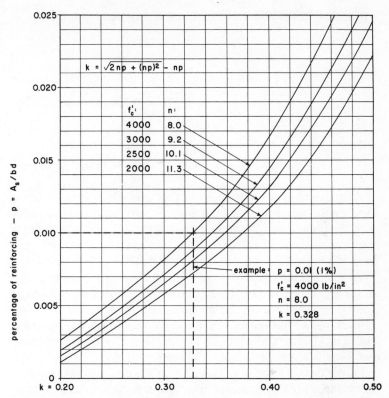

FIGURE 6.4 Flexural k factors for rectangular sections with tensile reinforcement only—as a function of p and n.

to be stressed beyond the capacity of the concrete to develop the required moment. Using formula (8),

$$M = Rbd^2 = 200 \text{ kip-ft } [271 \text{ kN-m}]$$

Then from Table 6.1, for f'_c of 3000 psi and f_s of 20 ksi,

$$R = 0.226 \text{ (in units of kip-in.) } [1554 \text{ in units of kN-m}]$$

Therefore,

$$M = 200 \times 12 = 0.226(bd^2), \text{ and } bd^2 = 10{,}619$$

(2) Various combinations of b and d may be found; for example,

$$b = 10 \text{ in.}, \quad d = \sqrt{\frac{10{,}619}{10}}$$

$$= 32.6 \text{ in. } [b = 0.254 \text{ m}, \quad d = 0.829 \text{ m}]$$

$$b = 15 \text{ in.}, \quad d = \sqrt{\frac{10{,}619}{15}}$$

$$= 26.6 \text{ in. } [b = 0.381 \text{ m}, \quad d = 0.677 \text{ m}]$$

Although they are not given in this example, there are often some considerations other than flexural behavior alone that influence the choice of specific dimensions for a beam. These situations are discussed in Chapter 7. If the beam is of the ordinary form shown in Fig. 6.5, the specified dimension is usually that given as h. Assuming the use of a No. 3 U-stirrup, a cover of 1.5 in. [38 mm], and an average-size reinforcing bar of 1-in. [25-mm] diameter (No. 8 bar), the design dimension d will be less than h by 2.375 in. [60 mm]. Lacking other considerations, we will assume a b of 15 in. [380 mm] and an h of 29 in. [740 mm], with the resulting d of $29 - 2.375 = 26.625$ in. [680 mm].

(3) We next use the specific value for d with formula (4) to find the required area of steel A_s. Since our selection is very close to the balanced section, we may use the value of j from Table 6.1. Thus

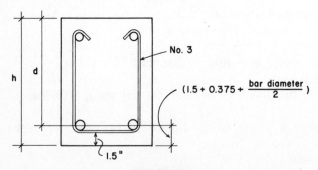

FIGURE 6.5

$$A_s = \frac{M}{f_s jd} = \frac{200 \times 12}{20 \times 0.872 \times 26.625} = 5.17 \text{ in.}^2 \text{ [3312 mm}^2\text{]}$$

Or using the formula for the definition of p and the balanced p value from Table 6.1,

$$A_s = pbd = 0.0129(15 \times 26.625) = 5.15 \text{ in.}^2 \text{ [3333 mm}^2\text{]}$$

(4) We next select a set of reinforcing bars to obtain this area. As with the beam dimensions, there are other concerns, as discussed in Chapter 7. For the purpose of our example, if we select bars all of a single size (see Table 2.2), the number required will be:

For No. 6 bars, $\dfrac{5.17}{0.44} = 11.75$, or 12 $\left[\dfrac{3312}{284} = 11.66\right]$

For No. 7 bars, $\dfrac{5.17}{0.60} = 8.62$, or 9 $\left[\dfrac{3312}{387} = 8.56\right]$

For No. 8 bars, $\dfrac{5.17}{0.79} = 6.54$, or 7 $\left[\dfrac{3312}{510} = 6.49\right]$

For No. 9 bars, $\dfrac{5.17}{1.00} = 5.17$, or 6 $\left[\dfrac{3312}{645} = 5.13\right]$

For No. 10 bars, $\dfrac{5.17}{1.27} = 4.07$, or 5 $\left[\dfrac{3312}{819} = 4.04\right]$

For No. 11 bars, $\dfrac{5.17}{1.56} = 3.31$, or 4 $\left[\dfrac{3312}{1006} = 3.29\right]$

In real design situations there are always various additional considerations that influence the choice of the reinforcing bars. One general desire is that of having the bars in a single layer, as this keeps the centroid of the steel as close as possible to the edge (bottom in this case) of the member, giving the greatest value for d with a given height of concrete section. With the section as shown in Fig. 6.5, a beam width of 15 in. will yield a net width of 11.25 in. inside the No. 3 stirrups. (Outside width of $15 - 2 \times 1.5$ cover and 2×0.375 stirrup diameter.) Applying the criteria for minimum spacing as described in Sec. 4.4, the required width for the various bar combinations can be determined. Two examples for this are shown in Fig. 6.6. It will be found that the four No. 11 bars are the only choice that will fit this beam width.

Example 2. A rectangular beam of concrete with f'_c of 3000 psi [20.7 MPa] and steel with f_s of 20 ksi [138 MPa] has dimensions of $b = 15$ in. [380 mm] and $h = 36$ in. [910 mm]. Find the area required for the steel reinforcing for a moment of 200 kip-ft [271 kN-m].

Solution: The first step in this case is to determine the balanced moment capacity of the beam with the given dimensions. If we assume the section to be as shown in Fig. 6.5, we may assume an approximate value for d to be h minus 2.5 in. [64 mm], or 33.5 in. [851 mm]. Then with the value for R from Table 6.1,

$$M = Rbd^2 = 0.226 \times 15 \times (33.5)^2 = 3804 \text{ kip-in.}$$

or $\qquad M = \dfrac{3804}{12} = 317 \text{ kip-ft}$

$$[M = 1554 \times 0.380 \times (0.850)^2 = 427 \text{ kN-m}]$$

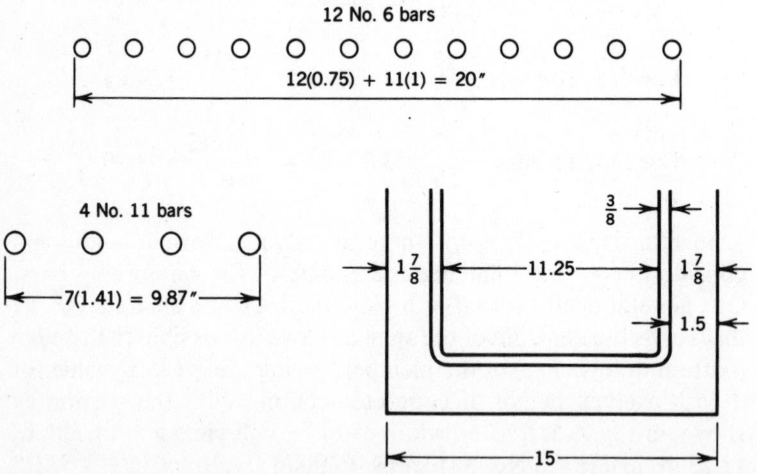

FIGURE 6.6 Consideration of beam width for proper spacing of a single layer of reinforcement.

Since this value is considerably larger than the required moment, it is thus established that the given section is larger than that required for a balanced stress condition. As a result, the concrete flexural stress will be lower than the limit of $0.45f_c'$, and the section is qualified as being under-reinforced, which is to say that the reinforcing required will be less than that required to produce a balanced section (with moment capacity of 317 kip-ft). In order to find the required area of steel, we use formula (4) just as we did in the preceding example. However, the true value for j in the formula will be something greater than that for the balanced section (0.872 from Table 6.1).

As the amount of reinforcing in the section decreases below the full amount required for a balanced section, the value of k decreases and the value of j increases. However, the range for j is small: from 0.872 up to something less than 1.0. A reasonable procedure is to assume a value for j, find the corresponding required area, and then perform an investigation to verify the assumed value for j, as follows. Assume $j = 0.90$. Then

$$A_2 = \frac{M}{f_s j d} = \frac{200 \times 12}{20 \times 0.90 \times 33.5} = 3.98 \text{ in.}^2 \ [2567 \text{ mm}^2]$$

and

$$p = \frac{A_s}{bd} = \frac{3.98}{15 \times 33.5} = 0.00792$$

Using this value for p in Fig. 6.4, we find $k = 0.313$. Using formula (6) we then determine j to be

$$j = 1 - \frac{k}{3} = 1 - \frac{0.313}{3} = 0.896$$

which is reasonably close to our assumption, so the computed area is adequate for design.

PROBLEM 6.4.A. A rectangular concrete beam has concrete with $f_c' = 3000$ psi [20.7 MPa] and steel reinforcing with $f_s = 20$ ksi [138 MPa]. Select the beam dimensions and reinforcing for a balanced section if the beam sustains a bending moment of 240 kip-ft [325 kN-m].

Problem 6.4.B. Find the area of steel reinforcing required and select the bars for the beam in Problem 6.4.A if the section dimensions are $b = 16$ in. and $d = 32$ in.

6.5 INVESTIGATION AND DESIGN: STRENGTH METHOD

Application of the working stress method consists of designing members to *work* in an adequate manner (without exceeding established stress limits) under actual service load conditions. The basic procedure in strength design is to design members to *fail;* thus the ultimate strength of the member at failure (called its design strength) is the only type of resistance considered. Safety in strength design is not provided by limiting stresses, as in the working stress method, but by using a factored design load (called the *required strength*) that is greater than the service load. The code establishes the value of the required strength, called U, as not less than

$$U = 1.4D + 1.7L$$

where D = effect of dead load

L = effect of live load

Other adjustment factors are provided when design conditions involve consideration of the effects of wind, earth pressure, differential settlement, creep, shrinkage, or temperature change.

The *design strength* of structural members (i.e., their *usable* ultimate strength) is determined by the application of assumptions and requirements given in the code and is further modified by the use of a *strength reduction factor* ϕ as follows:

ϕ = 0.90 for flexure, axial tension, and combinations of flexure and tension

= 0.75 for columns with spirals

= 0.70 for columns with ties

= 0.85 for shear and torsion

= 0.70 for compressive bearing

= 0.65 for flexure in plain (not reinforced) concrete

Thus while the formula for U may imply a relatively low safety factor, an additional margin of safety is provided by the stress-reduction factors.

Figure 6.7 shows the rectangular "stress block" that was described in Sec. 6.3 and is used for analysis of the rectangular section with tension reinforcing only by the strength method. This is the basis for investigation and design as provided for in the ACI Code.

The rectangular stress block is based on the assumption that a concrete stress of $0.85f'_c$ is uniformly distributed over the compression zone, which has dimensions equal to the beam width b and the distance a which locates a line parallel to and above the neutral axis. The value of a is determined from the expression $a = \beta_1 \times c$, where β_1 (beta one) is a factor that varies with the compressive strength of the concrete, and c is the distance from the extreme fiber to the neutral axis. For concrete having f'_c equal to or less than 4000 psi [27.6 MPa], the Code gives $a = 0.85\ c$.

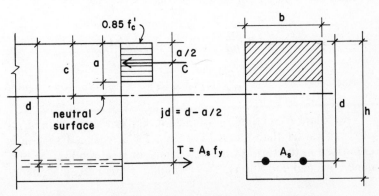

FIGURE 6.7 Bending resistance—rectangular stress block.

With the rectangular stress block, the magnitude of the compressive force in the concrete is expressed as

$$C = 0.85f_c' \times b \times a$$

and it acts at a distance of $a/2$ from the top of the beam. The arm of the resisting force couple then becomes $d - (a/2)$, and the developed resisting moment as governed by the concrete is

$$M_t = C\left(d - \frac{a}{2}\right) = 0.85f_c'ba \times \left(d - \frac{a}{2}\right) \tag{1}$$

With T expressed as $A_x \times f_y$, the developed moment as governed by the reinforcing is

$$M_t = T\left(d - \frac{a}{2}\right) = A_s f_y \left(d - \frac{a}{2}\right) \tag{2}$$

A formula for the dimension a of the stress block can be derived by equating the compression and tension forces; thus

$$0.85f_c'ba = A_s f_y \quad \text{and} \quad a = \frac{A_s f_y}{0.85f_c'b} \tag{3}$$

Expressing the area of steel in terms of a percentage ρ, the formula for a may be modified as follows:

$$\rho = \frac{A_s}{bd} \quad \text{or} \quad A_s = \rho bd$$

$$a = \frac{(\rho bd)f_y}{0.85f_c'b} = \frac{\rho d f_y}{0.85f_c'} \tag{4}$$

The balanced section for strength design is visualized in terms of strain rather than stress. The limit for a balanced section is expressed in the form of the percentage of steel required to produce balanced conditions. The formula for this percentage is

$$\rho_b = \frac{0.85f_c'}{f_y} \times \frac{87}{87 + f_y} \tag{5}$$

in which f'_c and f_y are in units of ksi. The ACI Code limits the percentage of steel to 75% of this balanced value in beams with tension reinforcing only.

Returning to the formula for the developed resisting moment, as expressed in terms of the steel, we see that a useful formula may be derived as follows:

$$M_t = A_s f_y \left(d - \frac{a}{2}\right) = (\rho bd)(f_y) \left(d - \frac{a}{2}\right)$$

$$= (\rho bd)(f_y)(d) \left(1 - \frac{1}{2}\frac{a}{d}\right)$$

$$= (bd^2) \left[\rho f_y \left(1 - \frac{1}{2}\frac{a}{d}\right)\right]$$

$$= Rbd^2 \tag{6}$$

where

$$R = \rho f_y \left(1 - \frac{1}{2}\frac{a}{d}\right) \tag{7}$$

With the reduction factor applied, the design moment for a section is limited to nine-tenths of the theoretical resisting moment.

Values for the balanced section factors—ρ, R, and a/d—are given in Table 6.2 for various combinations of f'_c and f_y. The balanced section, as discussed in the preceding section, is not necessarily a practical one for design. In most cases economy will be achieved by using less than the balanced reinforcing for a given concrete section. In special circumstances it may also be possible, or even desirable, to use compressive reinforcing in addition to tension reinforcing. Nevertheless, just as in the working stress method, the balanced section is often a useful reference when design is performed. The following example illustrates a procedure for the design of a simple rectangular beam section with tension reinforcing only.

Example 1. The service load bending moments on a beam are 58 kip-ft [78.6 kN-m] for dead load and 38 kip-ft [51.5 kN-m] for live

TABLE 6.2 Balanced Section Properties for Rectangular Concrete Sections with Tension Reinforcing Only: Strength Design[a]

f'_c		f_y = 40 ksi [276 MPa]					f_y = 60 ksi [414 MPa]				
		Balanced	Usable a/d	Usable	Usable R		Balanced	Usable a/d	Usable	Usable R	
psi	MPa	a/d	(75% Balance)	ρ	kip-in.	kN-m	a/d	(75% Balance)	ρ	kip-in.	kN-m
2000	13.79	0.5823	0.4367	0.0186	0.580	4000	0.5031	0.3773	0.0107	0.520	3600
2500	17.24	0.5823	0.4367	0.0232	0.725	5000	0.5031	0.3773	0.0137	0.650	4500
3000	20.69	0.5823	0.4367	0.0278	0.870	6000	0.5031	0.3773	0.0160	0.781	5400
4000	27.58	0.5823	0.4367	0.0371	1.161	8000	0.5031	0.3773	0.0214	1.041	7200
5000	34.48	0.5480	0.4110	0.0437	1.388	9600	0.4735	0.3551	0.0252	1.241	8600

[a] See Sec. 6.4 for derivation of formulas used to obtain table values.

load. The beam is 10 in. [254 mm] wide, f'_c is 4000 psi [27.6 MPa], and f_y is 60 ksi [414 MPa]. Determine the depth of the beam and the tensile reinforcing required.

Solution: (1) The first step is to determine the required moment, using the load factors. Thus

$$U = 1.4D + 1.7L$$

$$M_u = 1.4(M_{DL}) + 1.7(M_{LL})$$

$$= 1.4(58) + 1.7(38) = 145.8 \text{ kip-ft } [197.7 \text{ kN-m}]$$

(2) With the capacity reduction factor of 0.90 applied, the desired moment capacity of the section is determined as

$$M_t = \frac{M_u}{0.90} = \frac{145.8}{0.90} = 162 \text{ kip-ft} \quad \text{or} \quad 1944 \text{ kip-in. } [220 \text{ kN-m}]$$

(3) The maximum usable reinforcement ratio as given in Table 6.2 is $\rho = 0.0214$. If a balanced section is used, the required area of reinforcement may thus be determined from the relationship

$$A_s = \rho bd$$

While there is nothing especially desirable about a balanced section, it does represent the beam section with least depth if tension reinforcing only is used. We will therefore proceed to find the required balanced section for this example.

(4) To determine the required effective depth d, we use formula (6); thus

$$M_t = Rbd^2$$

With the value of $R = 1.041$ from Table 6.2,

$$M_t = 1944 = 1.041(10)(d)^2$$

and

$$d = \sqrt{\frac{1944}{1.041(10)}} = \sqrt{186.7} = 13.66 \text{ in.}$$

$$\left[d = \sqrt{\frac{220}{7200(0.254)}} = 0.347 \text{ m} \right]$$

(5) If this value is used for d, the required steel area may be found as

$$A_s = \rho bd = 0.0214(10)(13.66) = 2.92 \text{ in.}^2$$

The ACI Code requires a minimum ratio of reinforcing as follows:

$$\rho_{min} = \frac{200}{f_y} = \frac{200}{60,000} = 0.0033$$

which is clearly not critical for this example.

Selection of the actual beam dimensions and the actual number and size of reinforcing bars would involve various considerations, as discussed in Sec. 6.4. We will not complete the example in this case, since more complete design situations will be illustrated in the examples in the succeeding chapters.

If there are reasons, as there often are, for not selecting the least deep section with the greatest amount of reinforcing, a slightly different procedure must be used, as illustrated in the following example.

Example 2. Using the same data as in Example 1, find the reinforcing required if the desired beam section has $b = 10$ in. [254 mm] and $d = 18$ in. [457 mm].

Solution: The first two steps in this situation would be the same as in Example 1—to determine M_u and M_t. The next step would be to determine whether the given section is larger than, smaller than, or equal to a balanced section. Since this investigation has already been done in Example 1, we may observe that the 10×18 in. section is larger than a balanced section. Thus the actual value of a/d will be less than the balanced section value of 0.3773. The next step would then be as follows:

(4) Estimate a value for a/d—something smaller than the balanced value. For example, try $a/d = 0.25$. Then

$$a = 0.25d = 0.25(18) = 4.5 \text{ in. } [114 \text{ mm}]$$

With this assumed value for a, we may use formula (2) to find a value for A_s.

(5) Referring to Sec. 6.4 and Fig. 6.6,

$$A_s = \frac{M_t}{f_y(d - a/2)} = \frac{1944}{60(15.75)} = 2.057 \text{ in.}^2$$

(6) We next test to see if the estimate for a/d was close by finding a/d using formula (5) of Sec. 6.4. Thus

$$\rho = \frac{A_s}{bd} = \frac{2.057}{10(18)} = 0.0114$$

and

$$\frac{a}{d} = \rho \frac{f_y}{0.85f_c'} = 0.0114 \frac{60}{0.85(4)} = 0.202$$

$$a = 0.202(18) = 3.63 \text{ in.}, \quad d - \frac{a}{2} = 16.2 \text{ in.}$$

If we replace the value for $d = a/2$ that was used earlier with this new value, the required value of A_s will be slightly reduced. In this example, the correction will be only a few percent. If the first guess for a/d had been way off, it may justify a second run through steps 4, 5, and 6 to get closer to an exact answer.

PROBLEM 6.5.A,B,C. Using $f_c' = 3$ ksi [20.7 MPa] and $f_y = 50$ ksi [345 MPa], find the minimum depth required for a balanced section for the given data. Also find the area of reinforcement required if the depth chosen is 1.5 times that required for the balanced section. Use strength design methods.

	Moment Due to:				Beam Width	
	Dead Load		Live Load			
	kip-ft	kN-m	kip-ft	kN-m	(in.)	(mm)
A	40	54.2	20	27.1	12	305
B	80	108.5	40	54.2	15	381
C	100	135.6	50	67.8	18	457

6.6 BEAMS WITH COMPRESSIVE REINFORCEMENT

There are many situations in which steel reinforcing is used on both sides of the neutral axis in a beam. When this occurs, the steel on one side of the axis will be in tension and that on the other side in compression. Such a beam is referred to as a doubly reinforced beam or simply as a beam with compressive reinforcing (it being naturally assumed that there is also tensile reinforcing). Various situations involving such reinforcing have been discussed in the preceding sections. In summary, the most common occasions for such reinforcing include:

1. The desired resisting moment for the beam exceeds that for which the concrete alone is capable of developing the necessary compressive force.
2. Other functions of the section require the use of reinforcing on both sides of the beam. These include the need for bars to support U-stirrups and situations when torsion is a major concern.
3. It is desired to reduce deflections by increasing the stiffness of the compressive side of the beam. This is most significant for reduction of long-term creep deflections.
4. The combination of loading conditions on the structure result in reversal moments on the section; that is, the section must sometimes resist positive moment, and other times resist negative moment.

5. Anchorage requirements (for development of reinforcing) require that the bottom bars in a continuous beam be extended a significant distance into the supports.

The precise investigation and accurate design of doubly reinforced sections, whether performed by the working stress or by strength design methods, are quite complex and are beyond the scope of work in this book. The following discussion presents an approximation method that is adequate for preliminary design of a doubly reinforced section. For real design situations, this method may be used to establish a first trial design, which may then be more precisely investigated using more rigorous methods.

For the beam with double reinforcing, as shown in Fig. 6.8a, we consider the total resisting moment for the section to be the sum of the following two component moments.

M_1 (Fig. 6.8b) is comprised of a section with tension reinforcing only (A_{s1}). This section is subject to the usual procedures for design and investigation, as discussed in Secs. 6.4 and 6.5.

M_2 (Fig. 6.8c) is comprised of two opposed steel areas (A_{s2} and A_s') that function in simple moment couple action, similar to the flanges of a steel beam or the top and bottom chords of a truss.

Ordinarily, we expect that $A_{s2} = A_s'$, since the same grade of steel is usually used for both. However, there are two special considerations that must be made. The first involves the fact that A_{s2} is in tension, while A_s' is in compression. A_s' must therefore be dealt with in a manner similar to that for column reinforcing. This requires, among other things, that the compressive reinforcing be braced against buckling, using ties similar to those in a tied column.

The second consideration involves the distribution of stress and strain on the section. Referring to Fig. 6.8d, it may be observed that, under normal circumstances (kd less than $0.5d$), A_s' will be closer to the neutral axis than A_{s2}. Thus the stress in A_s' will be lower than that in A_{s2} if pure elastic conditions are assumed. However, it is common practice to assume steel to be doubly stiff when sharing stress with concrete in compression, due to shrinkage and creep effects. Thus, in translating from linear strain conditions to stress distribution, we use the relation $f_s'/2n$ (where $n = E_s/E_c$, as discussed in Sec. 2.6). Utilization of this relationship is illustrated in the following examples.

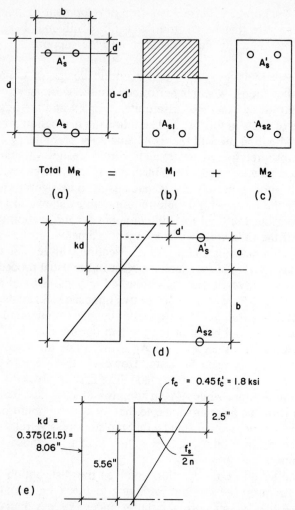

FIGURE 6.8 Basis for simplified analysis of a doubly reinforced beam.

Example 1. A concrete section with $b = 18$ in. [0.457 m] and $d = 21.5$ in. [0.546 m] is required to resist service load moments as follows: dead load moment = 150 kip-ft [203.4 kN-m], live load moment = 150 kip-ft [203.4 kN-m]. Using working stress methods, find the required reinforcing. Use $f'_c = 4$ ksi [27.6 MPa] and $f_y = 60$ ksi [414 MPa].

Solution: For the grade 60 reinforcing, we use an allowable stress of $f_s = 24$ ksi [165 MPa]. Then, using Table 6.1, find

$$n = 8, \quad k = 0.375, \quad j = 0.875, \quad p = 0.0141$$

$$R = 0.295 \text{ in kip-in. units } [2028 \text{ in kN-m units}]$$

Using the R value for the balanced section, the maximum resisting moment of the section is

$$M = Rbd^2 = \frac{0.295}{12} \times (18)(21.5)^2$$

$$= 205 \text{ kip-ft } [278 \text{ kN-m}]$$

This is M_1, as shown in Fig. 6.8*b*. Thus

$$M_2 = \text{total } M - M_1 = 300 - 205 = 95 \text{ kip-ft}$$
$$[407 - 278 = 129 \text{ kN-m}]$$

For M_1 the required reinforcing (A_{s1} in Fig. 6.8*b*) may be found as

$$A_{s1} = pbd = 0.0141 \times 18 \times 21.5 = 5.46 \text{ in.}^2 \text{ [3523 mm}^2]$$

and assuming that $f_s' = f_s$, we find A_s' and A_{s2} as follows:

$$M_2 = A_s' f_s'(d - d') = A_{s2} f_s(d - d')$$

$$A_s' = A_{s2} = \frac{M_2}{f_s(d - d')} = \frac{95 \times 12}{24 \times 19} = 2.50 \text{ in.}^2 \text{ [1613 mm}^2]$$

The total tension reinforcing is thus

$$A_s = A_{s1} + A_{s2} = 5.46 + 2.50 = 7.96 \text{ in.}^2 \text{ [5136 mm}^2]$$

For the compressive reinforcing, we must find the proper limit for f_s'. To do this, we assume the neutral axis of the section to be that for the balanced section, producing the situation that is shown in Fig. 6.8*e*. Based on this assumption, the limit for f_s' is found as follows:

$$\frac{f'_s}{2n} = \frac{5.56}{8.06}(0.45 \times 4) = 1.24 \text{ ksi}$$

$$f'_s = 2n \times 1.24 = 2 \times 8 \times 1.24 = 19.84 \text{ ksi [137 MPa]}$$

Since this is less than the limit of 24 ksi, we must use it to find A'_s; thus

$$A' = \frac{M_2}{f'_s(d - d')} = \frac{95 \times 12}{19.84 \times 19} = 3.02 \text{ in.}^2 \text{ [1948 mm}^2]$$

In practice, compressive reinforcement is often used even when the section is theoretically capable of developing the necessary resisting moment with tension reinforcement only. This calls for a somewhat different procedure, as is illustrated in the following example.

Example 2. Design the beam in Example 1 using strength design methods.

Solution: We first find the design moment in the usual manner.

$$M_u = 1.4M_d + 1.7M_l = 1.4(150) + 1.7(150)$$

$$= 465 \text{ kip-ft [631 kN-m]}$$

$$M_t = \frac{M_u}{\phi} = \frac{465}{0.9} = 517 \text{ kip-ft [701 kN-m]}$$

As a point of reference, we next determine the maximum resisting moment for the section with tension reinforcing only. Thus, using Table 6.2, we find that

$$R = 1.041, \quad p = 0.0214, \quad \frac{a}{d} = 0.3773$$

and

$$M = Rbd^2 = \frac{1.041}{12}(18)(21.5)^2$$

$$= 722 \text{ kip-ft [979 kN-m]}$$

This indicates that the section could actually function without compressive reinforcing. However, we will assume that there are compelling reasons for having some compressive reinforcing although its *amount* (magnitude of A'_s) becomes somewhat arbitrary. As a rough guide, we suggest a trial design with A'_s approximately one-third of A_s. On the basis of the previous computation, we know that the value for A_s will be less than that required for the full maximum resisting moment. That is, A_s will be less than

$$A_s = pbd = 0.0214(18)(21.5) = 8.28 \text{ in.}^2 \text{ [5342 mm}^2]$$

For a trial design, we choose compressive reinforcing consisting of two No. 9 bars, with $A'_s = 2.0$ in.^{2} [1290 mm^2]. With this reinforcing we may now compute a value for M_2, but to do so we must first establish a value for f'_s, the usable stress in the compressive reinforcing. For an approximate design, we may use the relationship shown in Fig. 6.9, in which we visualize the limit for f'_s to be $2n$ times the maximum stress of $0.85 f'_c$ in the concrete. Thus

$$f'_s = 2n(0.85f'_c) = 2(8)[0.85(4)]$$

$$= 54.4 \text{ ksi [375 MPa]}$$

Since this value is less than the limiting yield strength, we use it to find M_2, thus

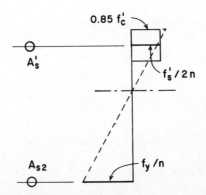

FIGURE 6.9 Stress and strain in the doubly reinforced beam.

$$M_2 = A'_s f'_s (d - d')$$

$$= 2.0(54.4)(19)(\tfrac{1}{12}) = 172 \text{ kip-ft } [233 \text{ kN-m}]$$

A_{s2} will have a value different from A'_s, since the value for stress for A_{s2} will be the full yield stress of 60 ksi. Thus

$$M_2 = A_{s2} f_y (d - d') = 172 \text{ kip-ft}$$

$$A_{s2} = \frac{172(12)}{60(19)} = 1.81 \text{ in.}^2 \ [1168 \text{ mm}^2]$$

With the value of M_2 established, we now find the required value for M_1. Thus

$$M_1 = M_t - M_2 = 517 - 172 = 345 \text{ kip-ft}$$

$$[701 - 233 = 468 \text{ kN-m}]$$

To find the required value for A_{s1}, we use the usual procedure for a section with tension reinforcing only, as described in Sec. 6.4. Since the required value for M_1 is almost half of the maximum resisting moment (722 kip-ft, as computed previously), we may assume that a/d will be considerably smaller than the table value of 0.377. For a first guess, try

$$\frac{a}{d} = 0.20, \qquad a = 0.20(21.5) = 4.3 \text{ in.}$$

Rounding this off to 4 in., we find

$$A_{s1} = \frac{M_1}{f_y(d - a/2)} = \frac{345(12)}{60(19.5)} = 3.53 \text{ in.}^2 \ [2277 \text{ mm}^2]$$

With this area of steel, $p = 3.53/[18(21.5)] = 0.00912$ and

$$a = pd \frac{f_y}{0.85f'_c} = 0.00912(21.5) \frac{60}{0.85(4)}$$

$$= 3.46 \text{ in.}$$

For a second try, guess $a = 3.4$ in. Then

$$A_{s1} = \frac{345(12)}{60(19.8)} = 3.48 \text{ in.}^2$$

and

$$\text{total } A_s = A_{s1} + A_{s2} = 3.48 + 1.81 = 5.29 \text{ in.}^2$$

With these computations completed, we now make a choice of reinforcing for the section as follows.

Compressive Reinforcing: two No. 9 bars, $A'_s = 2.0$ in.^{2} [1290 mm^2].

Tensile Reinforcing: two No. 10 + two No. 11 bars, $A_s = 5.66$ in.^{2} [3652 mm^2].

The following example illustrates a procedure that may be used with the working stress method, when the required resisting moment is less than the balanced section limiting moment. It is generally similar to the procedure used with strength design in Example 2.

Example 3. Design a doubly reinforced section by the working stress method for a moment of 180 kip-ft [244 kN-m]. Use the section dimensions and data given in Example 1.

Solution: The first step is to investigate the section for its balanced stress limiting moment, as was done in Example 1. This will show that the required moment is less than the balanced moment limit, and that the section could function without compressive reinforcing. Again, we assume that compressive reinforcing is desired, so we assume an arbitrary amount for A'_s and proceed as in Example 2. We make a first guess for the total tension reinforcing as

$$A_s = \frac{M}{f_s(0.9d)} = \frac{180(12)}{24[0.9(21.5)]} = 4.65 \text{ in.}^2 \text{ [3000 mm}^2\text{]}$$

Try

$$A_s' = \tfrac{1}{3}A_s = \tfrac{1}{3}(4.65) = 1.55 \text{ in.}^2 \text{ [1000 mm}^2]$$

Choose two No. 8 bars,

$$\text{actual } A_s' = 1.58 \text{ in.}^2 \text{ [1019 mm}^2]$$

Thus

$$A_{s1} = A_s - A_s' = 4.65 - 1.58 = 3.07 \text{ in.}^2 \text{ [1981 mm}^2]$$

Using A_{s1} for a rectangular section with tension reinforcing only (see Sec. 6.4).

$$p = \frac{3.07}{18(21.5)} = 0.0079$$

Then, from Fig. 6.4, we find $k = 0.30, j = 0.90$.

Using these values for the section, and the formula involving the concrete stress in compression from Sec. 6.4, we find

$$f_c = \frac{2M_1}{kjbd^2} = \frac{2(120)(12)}{0.3(0.9)(18)(21.5)^2}$$
$$= 1.28 \text{ ksi [8.83 MPa]}$$

With this value for the maximum concrete stress and the value of 0.30 for k, the distribution of compressive stress will be as shown in Fig. 6.10. From this, we determine the limiting value for f_s' as follows:

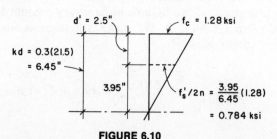

FIGURE 6.10

$$\frac{f'_s}{2n} = \frac{3.95}{6.45}(1.28) = 0.784 \text{ ksi}$$

$$f'_s = 2n(0.784) = 2(8)(0.784) = 12.5 \text{ ksi } [86.2 \text{ MPa}]$$

Since this is lower than f_s, we use it to find the limiting value for M_2. Thus

$$M_2 = A'_s f'_s (d - d')$$

$$= 1.58(12.5)(19)(\tfrac{1}{12}) = 31 \text{ kip-ft } [42 \text{ kN-m}]$$

To find A_{s2} we use this moment with the full value of $f_s = 24$ ksi. Thus

$$A_{s2} = \frac{M_2}{f_s(d - d')} = \frac{31(12)}{24(19.0)} = 0.82 \text{ in.}^2 \text{ [529 mm}^2\text{]}$$

To find A_{s1}, we determine that

$$M_1 = \text{total } M - M_2 = 180 - 31$$

$$= 149 \text{ kip-ft } [202 \text{ kN-m}]$$

$$A_{s1} = \frac{M_1}{f_s jd} = \frac{149(12)}{24(0.9)(21.5)} = 3.85 \text{ in.}^2 \text{ [2484 MPa]}$$

Then the total tension reinforcing is found as

$$A_s = A_{s1} + A_{s2} = 3.85 + 0.82$$

$$= 4.67 \text{ in.}^2 \text{ [3013 mm}^2\text{]}$$

PROBLEM 6.6.A. A concrete section with $b = 16$ in. [0.406 m] and $d = 19.5$ in. [0.495 m] is required to resist service load moments as follows: dead load moment = 120 kip-ft [163 kN-m], live load moment = 110 kip-ft [136 kN-m]. Using working stress methods, find the required reinforcing. Use $f'_c = 4$ ksi [27.6 MPa] and grade 60 bars with $f_y = 60$ ksi [414 MPa] and $f_s = 24$ ksi [165 MPa].

PROBLEM 6.6.B. Using strength methods, select the reinforcement for the beam in Problem 6.6.A. Use compressive reinforcement with approximately one-third the area of the tensile reinforcement.

PROBLEM 6.6.C. Using the working stress method, find the reinforcement required for the beam in Problem 6.6.A if the effective beam depth is 30 in. Use compressive reinforcement with approximately one-third the area of the tensile reinforcement.

6.7 T-BEAMS

When a floor slab and its supporting beams are poured at the same time, the result is a monolithic construction in which a portion of the slab on each side of the beam serves as the flange of a T-beam. The part of the section that projects below the slab is called the web or stem of the T-beam. This type of beam is shown in Fig. 6.11a. For positive moment, the flange is in compression and

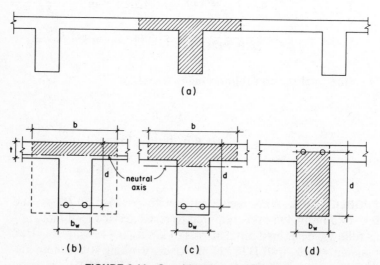

FIGURE 6.11 Considerations for T-beams.

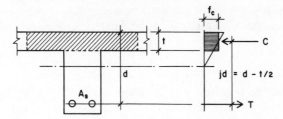

FIGURE 6.12 Basis for simplified analysis of a T-beam.

there is ample concrete to resist compressive stresses, as shown in Fig. 6.11*b* or *c*. However, in a continuous beam, there are negative bending moments over the supports, and the flange here is in the tension stress zone with compression in the web.

It is important to remember that only the area formed by the width of the web b_w and the effective depth d is to be considered in computing resistance to shear and to bending moment over the supports. This is the hatched area shown in Fig. 6.11*d*.

The effective flange width to be used in the design of symmetrical T-beams is limited to one-fourth the span length of the beam. In addition, the overhanging width of the flange on either side of the web is limited to eight times the thickness of the slab or one-half the clear distance to the next beam.

In monolithic construction with beams and one-way solid slabs, the effective flange area of the T-beams is usually quite capable of resisting the compressive stresses caused by positive bending moments. With a large flange area, as shown in Fig. 6.11*a*, the neutral axis of the section usually occurs quite high in the beam web, resulting in only minor compressive stresses in the web. If the compression developed in the web is ignored, the net compression force may be considered to be located at the centroid of the trapezoidal stress zone that represents the stress distribution in the flange. On this basis, the compression force is located at something less than $t/2$ from the top of the beam.

An approximate analysis of the T-section by the working stress method that avoids the need to find the location of the neutral axis and the centroid of the trapezoidal stress zone, consists of the following steps.

1. Ignore compression in the web and assume a constant value for compressive stress in the flange (see Fig. 6.12). Thus

$$jd = d - \frac{t}{2}$$

2. Find the required steel area as

$$A_s = \frac{M}{f_s jd} = \frac{M}{f_s(d - t/2)}$$

3. Check the compressive stress in the concrete as

$$f_c = \frac{C}{b_f t}, \quad \text{where } C = \frac{M}{jd} = \frac{M}{d - t/2}$$

The actual value of maximum compressive stress will be slightly higher, but will not be critical if this computed value is significantly less than the limit of $0.45f'_c$.

The following example illustrates the use of this procedure. It assumes a typical design situation in which the dimensions of the section (b_f, b_w, d, and t) are all predetermined by other design considerations and the design of the T-section is reduced to the requirement to determine the area of tension reinforcing.

Example 1. A T-section is to be used for a beam to resist positive moment. The following data is given: beam span = 18 ft [5.49 m], beams are 9 ft [2.74 m] center to center, slab thickness is 4 in. [0.102 m], beam stem dimensions are b_w = 15 in. [0.381 m] and d = 22 in. [0.559 m], f'_c = 4 ksi [27.6 MPa], f_y = 60 ksi [414 MPa], f_s = 24 ksi [165 MPa]. Find the required area and pick reinforcing bars for a dead load moment of 100 kip-ft [136 kN-m] plus a live load moment of 100 kip-ft [136 kN-m].

Solution: Using working stress design with the approximate method described previously:

(1) Determine the effective flange width (necessary only for a check on the concrete stress). The maximum value for the flange width is

$$b_f = \frac{\text{span}}{4} = \frac{18(12)}{4} = 54 \text{ in. } [1.37 \text{ m}]$$

or

$$b_f = \text{center-to-center beam spacing}$$
$$= 9(12) = 108 \text{ in. } [2.74 \text{ m}]$$

or

$$b_f = \text{beam stem width plus 16 times the slab thickness}$$
$$= 15 + 16(4) = 79 \text{ in. } [2.01 \text{ m}]$$

The limiting value is therefore 54 in. [1.37 m].
(2) Find the required steel area

$$A_s = \frac{M}{f_s(d - t/2)} = \frac{200(12)}{24(22 - 4/2)}$$
$$= 5.00 \text{ in.}^2 [3364 \text{ mm}^2]$$

(3) Pick bars using Table 6.3 which incorporates consideration for the adequacy of the stem width. From the table: Choose five No. 9 bars, actual $A_s = 5.00$ in.^{2} From Table 7.1 required width for five No. 9 bars is 14 in., less than the 15 in. provided.

TABLE 6.3 Options for the T-Beam Reinforcement with Required Area of 5.00 in.^{2}

Bar Size	No. of Bars	Actual Area (in.^{2})	Width Required[a] (in.)
7	9	5.40	too wide
8	7	5.53	too wide
9	5	5.00	14
10	4	5.08	13
11	4	6.28	14

[a] From Table 7.1.

(4) Check the concrete stress.

$$C = \frac{M}{jd} = \frac{200(12)}{20} = 120 \text{ kips [535 kN]}$$

$$f_c = \frac{C}{b_f t} = \frac{120}{54(4)} = 0.556 \text{ ksi [3.83 MPa]}$$

$$\text{limiting stress} = 0.45f_c' = 0.45(4)$$
$$= 1.8 \text{ ksi [12.4 MPa]}$$

Thus compressive stress in the flange is clearly not critical.

In a real design situation, of course, consideration would have to be given to problems of shear and possibly to problems of development lengths for the bars.

When using strength design methods for T-sections, we recommend a procedure similar to that described for the working stress method. This method and procedure assumes that the flange area of the T is so large that the concrete stress never gets up to its ultimate limit before the yield stress develops in the reinforcing. The following example illustrates the procedure.

Example 2. Perform the design for the beam described in Example 1 using strength design methods.

Solution: (1) As in Example 1, effective $b_f = 54$ in. [1.37 m].

(2) Design moment is found as

$$M_u = 1.4M_d + 1.7\,M_l$$
$$= 1.4(100 + 1.7(100)$$
$$= 310 \text{ kip-ft [420 kN-m]}$$

(3) Required design strength is found as

$$M_t = \frac{M_u}{0.90} = \frac{310}{0.90} = 345 \text{ kip-ft [467 kN-m]}$$

(4) Assuming the location of the net compression force to be at the center of the flange area, as described for the working stress method, the steel area is found as

$$A_s = \frac{M}{f_y(d - t/2)} = \frac{345(12)}{60(20)}$$

$$= 3.45 \text{ in.}^2 \ [2221 \text{ mm}^2]$$

(5) A possible choice for the reinforcing is two No. 8 bars plus two No. 9 bars, providing an actual area of 3.58 in.^{2} [2310 mm^2]. Table 7.1 shows that a minimum width for four No. 9 bars is only 12 in., so the stem width is more than adequate for bar spacing.

(6) Assuming the steel bars to be stressed to the yield point, the average stress in the flange would be as follows.

$$C = T = f_y A_s = 60(3.58)$$

$$= 215 \text{ kips} \ [956 \text{ kN}]$$

$$f_c = \frac{C}{b_f t} = \frac{215}{54(4)} = 0.995 \text{ ksi} \ [6.86 \text{ MPa}]$$

which is considerably lower than f'_c.

The examples in this section illustrate procedures that are reasonably adequate for beams that occur in ordinary beam and slab construction. When special T-sections occur with thin flanges (t less than $d/8$ or so) or narrow effective flange widths (b_f less than three times b_w or so), these methods may not be valid. In such cases more accurate investigation should be performed, using the requirements of the ACI Code.

PROBLEM 6.7.A. Find the area of steel reinforcement required for a concrete T-beam for the following data: allowable $f_s = 20$ ksi [138 MPa], $d = 28$ in. [711 mm], $t = 6$ in. [152 mm], $b' = 16$ in. [406 mm], and the section sustains a bending moment of 240 kip-ft [325 kN-m].

6.8 SPANNING SLABS

Concrete slabs are frequently used as spanning roof or floor decks, often occurring in monolithic, cast-in-place slab and beam framing systems. There are generally two basic types of slabs: one-way spanning and two-way spanning slabs. The spanning condition is not so much determined by the slab as by its support conditions. The two-way spanning slab is discussed in Sec. 7.6. As part of a general framing system, the one-way spanning slab is discussed in Sec. 7.3. The following discussion relates to the design of one-way solid slabs using procedures developed for the design of rectangular beams.

Solid slabs are usually designed by considering the slab to consist of a series of 12-in.-wide planks. Thus the procedure consists of simply designing a beam section with a predetermined width of 12 in. Once the depth of the slab is established, the required area of steel is determined, specified as the number of square inches of steel required per foot of slab width.

Reinforcing bars are selected from a limited range of sizes, appropriate to the slab thickness. For thin slabs (4 to 6 in. thick) bars may be of a size from No. 3 to No. 6 or so (nominal diameters from $\frac{3}{8}$ to $\frac{3}{4}$ in.). The bar size selection is related to the bar spacing, the combination resulting in the amount of reinforcing in terms of so many square inches per one foot unit of slab width. Spacing is limited by code regulation to a maximum of three times the slab thickness. There is no minimum spacing, other than that required for proper placing of the concrete; however, a very close spacing indicates a very large number of bars, making for laborious installation.

Every slab must be provided with two-way reinforcement, regardless of its structural functions. This is to satisfy requirements for shrinkage and temperature effects, as discussed in Chapter 4. The amount of this minimum reinforcement is specified as a percentage p of the gross cross-sectional area of the concrete, as follows:

1. For slabs reinforced with grade 40 or grade 50 bars:

$$p = \frac{A_s}{bt} = 0.0020 \quad \text{or} \quad 0.2\%$$

2. For slabs reinforced with grade 60 bars:

$$p = \frac{A_s}{bt} = 0.0018 \quad \text{or} \quad 0.18\%$$

Center-to-center spacing of this minimum reinforcement must not be greater than five times the slab thickness or 18 in.

Minimum cover for slab reinforcement is normally 0.75 in., although exposure conditions or need for a high fire rating may require additional cover. For a thin slab reinforced with large bars, there will be a considerable difference between the slab thickness and the effective depth—t and d, as shown in Fig. 6.13. Thus the practical efficiency of the slab in flexural resistance decreases rapidly as the slab thickness is decreased. For this and other reasons, very thin slabs (less than 4 in. thick) are often reinforced with wire fabric rather than sets of loose bars.

Shear reinforcement is seldom used in one-way slabs, and consequently the maximum unit shear stress in the concrete must be kept within the limit for the concrete without reinforcement. This is usually not a concern, as unit shear is usually low in one-way slabs, except for exceptionally high loadings.

Table 6.4 gives data that are useful in slab design, as demonstrated in the following example. Table values indicate the average amount of steel area per foot of slab width provided by various combinations of bar size and spacing. Table entries are determined as follows:

$$A_s/\text{ft} = (\text{bar area}) \frac{12}{\text{bar spacing}}$$

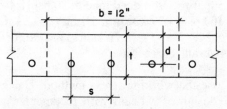

FIGURE 6.13 Reference for slab design.

TABLE 6.4 Areas of Bars in Reinforced Concrete Slabs per Foot of Width

Bar Spacing (in.)	Areas of Bars (in.^{2}/ft)									
	No. 2	No. 3	No. 4	No. 5	No. 6	No. 7	No. 8	No. 9	No. 10	No. 11
3	0.20	0.44	0.80	1.24	1.76	2.40	3.16	4.00		
3.5	0.17	0.38	0.69	1.06	1.51	2.06	2.71	3.43	4.35	
4	0.15	0.33	0.60	0.93	1.32	1.80	2.37	3.00	3.81	4.68
4.5	0.13	0.29	0.53	0.83	1.17	1.60	2.11	2.67	3.39	4.16
5	0.12	0.26	0.48	0.74	1.06	1.44	1.89	2.40	3.05	3.74
5.5	0.11	0.24	0.44	0.68	0.96	1.31	1.72	2.18	2.77	3.40
6	0.10	0.22	0.40	0.62	0.88	1.20	1.58	2.00	2.54	3.12
7	0.08	0.19	0.34	0.53	0.75	1.03	1.35	1.71	2.18	2.67
8	0.07	0.16	0.30	0.46	0.66	0.90	1.18	1.50	1.90	2.34
9	0.07	0.15	0.27	0.41	0.59	0.80	1.05	1.33	1.69	2.08
10	0.06	0.13	0.24	0.37	0.53	0.72	0.95	1.20	1.52	1.87
11	0.05	0.12	0.22	0.34	0.48	0.65	0.86	1.09	1.38	1.70
12	0.05	0.11	0.20	0.31	0.44	0.60	0.79	1.00	1.27	1.56
13	0.05	0.10	0.18	0.29	0.40	0.55	0.73	0.92	1.17	1.44
14	0.04	0.09	0.17	0.27	0.38	0.51	0.68	0.86	1.09	1.34
15	0.04	0.09	0.16	0.25	0.35	0.48	0.63	0.80	1.01	1.25
16	0.04	0.08	0.15	0.23	0.33	0.45	0.59	0.75	0.95	1.17
18	0.03	0.07	0.13	0.21	0.29	0.40	0.53	0.67	0.85	1.04
24	0.02	0.05	0.10	0.15	0.22	0.30	0.39	0.50	0.63	0.78

Thus for No. 5 bars at 8-in. centers,

$$A_s/\text{ft} = (0.31)\frac{12}{8} = 0.465 \text{ in.}^2/\text{ft}$$

It may be observed that the table entry for this combination is rounded off to a value of 0.46 in.^{2}/ft.

Example 1. A one-way solid concrete slab is to be used for a simple span of 14 ft [4.27 m]. In addition to its own weight, the slab carries a superimposed dead load of 30 psf [1.44 kN/m^2] and a live load of 100 psf [4.79 kN/m^2]. Using $f'_c = 3$ ksi [20.7 MPa], $f_y = 40$ ksi [276 MPa], and $f_s = 20$ ksi [138 MPa], design the slab for minimum overall thickness.

Solution: Choose the working stress method. Using the general procedure for design of a beam with rectangular section (Sec.

6.4), we first determine the required slab thickness. Thus for deflection, from Table 7.2,

$$\text{minimum } t = \frac{L}{25} = \frac{14(12)}{25} = 6.72 \text{ in. [171 mm]}$$

For flexure we first determine the maximum bending moment. The loading must include the weight of the slab, for which we use the thickness required for deflection as a first estimate. Assuming a 7-in. [178-mm]-thick slab, then slab weight is $\frac{7}{12}(150 \text{ pcf}) = 87.5$ psf, say 88 psf and total load is 100 psf LL + 118 psf DL = 218 psf.

The maximum bending moment for a 12-in.-wide design strip of the slab thus becomes

$$M = \frac{wL^2}{8} = \frac{218(14)^2}{8} = 5341 \text{ ft-lb [7.24 kN-m]}$$

For minimum slab thickness, we consider the use of a balanced section, for which Table 6.1 yields the following properties:

$$j = 0.872, \qquad p = 0.0129, \qquad R = 0.226$$

Then

$$bd^2 = \frac{M}{R} = \frac{5.341(12)}{0.226} = 284 \text{ in.}^3$$

and since b is the 12-in. design strip width,

$$d = \sqrt{\frac{284}{12}} = \sqrt{23.7} = 4.86 \text{ in. [123 mm]}$$

Assuming an average bar size of a No. 6 ($\frac{3}{4}$ in. nominal diameter) and cover of $\frac{3}{4}$ in., the minimum required slab thickness based on flexure becomes

$$t = 4.86 + \frac{0.75}{2} + 0.75 = 5.985 \text{ in.}$$

$$\left[123 + \frac{19}{2} + 19 = 152 \text{ mm} \right]$$

We thus observe that the deflection limitation controls in this situation, and the minimum overall thickness is the 6.72-in. dimension. If we continue to use the 7-in. overall thickness, the actual effective depth with a No. 6 bar will be

$$d = 7.0 - 1.125 = 5.875 \text{ in.}$$

Since this d is larger than that required for a balanced section, the value for j will be slightly larger than 0.872, as found from Table 6.1. Let us assume a value of 0.9 for j and determine the required area of reinforcement as

$$A_s = \frac{M}{f_s jd} = \frac{5.341(12)}{20(0.9)(5.875)} = 0.606 \text{ in.}^2$$

From Table 6.4, we find that the bar combinations shown in Table 6.5 will satisfy this requirement.

The ACI Code requires a maximum spacing of three times the slab thickness (21 in. in this case or 18 in., whichever is smaller). Minimum spacing is largely a matter of the designer's judgment. Many designers consider a minimum practical spacing to be one approximately equal to the slab thickness. Within these limits, any of the bar size and spacing combinations listed are adequate.

TABLE 6.5 Alternatives for the Slab Reinforcement

Bar Size	Spacing from Center to Center (in.)	Average A_s in a 12-in. Width
5	6	0.61
6	8.5	0.62
7	12	0.60
8	15	0.63

As described previously, the ACI Code requires a minimum reinforcement for shrinkage and temperature effects to be placed in the direction perpendicular to the flexural reinforcement. With the grade 40 bars in this example, the minimum percentage of this steel is 0.0020, and the steel area required for a 12-in. strip thus becomes

$$A_s = p(bt) = 0.0020[12(7)] = 0.168 \text{ in.}^2$$

From Table 6.4, we find that this requirement can be satisfied with No. 3 bars at 8-in. centers or No. 4 bars at 14-in. centers. Both of these spacings are well below the maximum of five times the slab thickness or 18 in.

Although simply supported single slabs are sometimes encountered, the majority of slabs used in building construction are continuous through multiple spans. An example of the design of such a slab is given in Chapter 7.

Strength Design Method

Strength design procedures for the slab are essentially the same as for the rectangular beam, as described in Sec. 6.5. In most cases, slab sections will be reinforced with steel areas well below those for a balanced section, so the procedure for a so-called under-reinforced section should be used. If the procedure illustrated in Sec. 6.5 is used for this example, it will be found that the required steel area is approximately 15% less than that required from the working stress method computations.

PROBLEM 6.8.A. A one-way solid concrete slab is to be used for a simple span of 16 ft [4.88 m]. In addition to its own weight, the slab carries a superimposed dead load of 60 psf [2.87 kN/m²] and a live load of 75 psf [3.59 kN/m²]. Using the working stress method with $f'_c = 3$ ksi [20.7 MPa], $f_y = 40$ ksi [276 MPa], and $f_s = 20$ ksi [138 MPa], design the slab for minimum overall thickness.

PROBLEM 6.8.B. Using the data from Problem 6.8.A, design the slab using strength design methods.

6.9 SHEAR IN BEAMS AND SLABS

There are many situations in concrete structures that involve the development of shear. In most cases the shear effect itself is not the major concern but rather, the diagonal tension that accompanies the shear action. The material in this chapter deals primarily with the situation of shear in slabs and beams. Shear conditions in footings, walls, and columns are dealt with in other chapters that deal generally with those elements.

General Concerns for Shear

The most common situations involving shear in concrete structures are shown in Fig. 6.14. Shear in beams (Fig. 6.14a) is ordinarily critical near the supports, where the shear force is greatest.

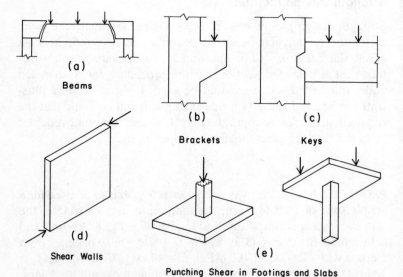

FIGURE 6.14 Situations involving shear in concrete structures.

In short brackets (Fig. 6.14*b*) and keys (Fig. 6.14*c*) the shear action is essentially a direct slicing effect. Punching shear, also called peripheral shear, occurs in column footings and in slabs that are directly supported on columns (Fig. 6.14*e*). When walls are used as bracing elements for shear forces that are parallel to the wall surface (called shear walls), they must develop resistance to the direct shear effect that is similar to that in a bracket.

In all of these situations consideration must be given to the shear effect and the resulting shear stresses. Both the magnitude and the direction of the shear stresses must be considered. In many cases, however, the shear effect occurs in combination with other effects, such as bending moment, axial tension, or axial compression. In combined force situations the resulting net combined stress situations must be considered.

Shear in Beams

From general consideration of shear effects, as developed in the science of mechanics of materials, the following observations can be made:

1. Shear is an ever-present phenomenon, produced directly by slicing actions, by lateral loading in beams, and on oblique sections in tension and compression members.

2. Shear forces produce shear stress in the plane of the force and equal unit shear stresses in planes that are perpendicular to the shear force.

3. Diagonal stresses of tension and compression, having magnitudes equal to that of the shear stress, are produced in directions of 45° from the plane of the shear force. ·

4. Direct slicing shear force produces a constant magnitude shear stress on affected sections, but beam shear action produces shear stress that varies on the affected sections, having magnitude of zero at the edges of the section and a maximum value at the centroidal neutral axis of the section.

In the discussions that follow it is assumed that the reader has a general familiarity with these relationships. Let us consider the case of a simple beam with uniformly distributed load and end supports that provide only vertical resistance (no moment restraint). The distribution of internal shear and bending moment are as shown in Fig. 6.15a. For flexural resistance, it is necessary to provide longitudinal reinforcing bars near the bottom of the beam. These bars are oriented for primary effectiveness in resistance to tension stresses that develop on a vertical (90°) plane (which is the case at the center of the span, where the bending moment is maximum and the shear approaches zero).

Under the combined effects of shear and bending, the beam tends to develop tension cracks as shown in Fig. 6.15b. Near the center of the span, where the bending is predominant and the shear approaches zero, these cracks approach 90°. Near the support, however, where the shear predominates and bending approaches zero, the critical tension stress plane approaches 45°, and the horizontal bars are only partly effective in resisting the cracking.

For beams, the most common form of shear reinforcement consists of a series of U-shaped bent bars (Fig. 6.15d), placed vertically and spaced along the beam span, as shown in Fig. 6.15c. These bars are intended to provide a vertical component of resistance, working in conjunction with the horizontal resistance provided by the flexural reinforcement. In order to develop tension near the support face, the horizontal bars must be bonded to the concrete beyond the point where the stress is developed. Where the beam ends extend only a short distance over the support (a common situation), it is often necessary to bend or hook the bars, as shown in Fig. 6.15.

The simple span beam and the rectangular section shown in Fig. 6.15 occur only infrequently in building structures. The most common case is that of the beam section shown in Fig. 6.16a, which occurs when a beam is poured monolithically with a supported concrete slab. In addition, these beams normally occur in continuous spans with negative moments at the supports. Thus the stress in the beam near the support is as shown in Fig. 6.16a, with the negative moment producing compressive flexural stress in the bottom of the beam stem. This is substantially different from the case of the simple beam, where the moment approaches zero near the support.

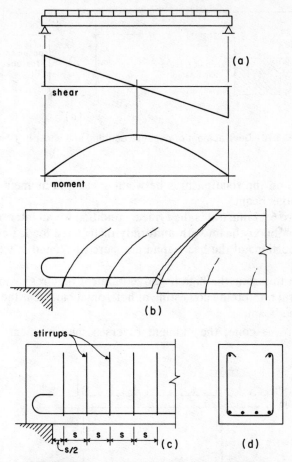

FIGURE 6.15 Consideration of shear in beams.

For the purpose of shear resistance, the continuous, T-shaped beam is considered to consist of the section indicated in Fig. 6.16*b*. The effect of the slab is ignored, and the section is considered to be a simple rectangular one. Thus for shear design, there is little difference between the simple span beam and the continuous beam, except for the effect of the continuity on the distribution of shear along the beam span. It is important, however, to

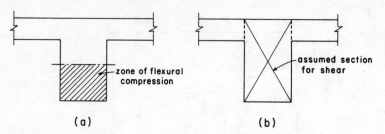

zone of flexural compression

assumed section for shear

(a)   (b)

FIGURE 6.16 Development of negative bending and shear in T-beams.

understand the relationships between shear and moment in the continuous beam.

Figure 6.17 illustrates the typical condition for an interior span of a continuous beam with uniformly distributed load. Referring to the portions of the beam span numbered 1, 2, and 3, we note:

1. In this zone the high negative moment requires major flexural reinforcing consisting of horizontal bars near the top of the beam.
2. In this zone, the moment reverses sign; moment magni-

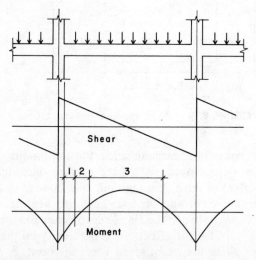

Shear

Moment

FIGURE 6.17 Shear and bending in continuous beams.

tudes are low; and, if shear stress is high, the design for shear is a predominant concern.

3. In this zone, shear consideration is minor and the predominant concern is for positive moment requiring major flexural reinforcing in the bottom of the beam.

Vertical U-shaped stirrups, similar to those shown in Fig. 6.18a, may be used in the T-shaped beam. An alternate detail for the U-shaped stirrup is shown in Fig. 6.18b, in which the top hooks are turned outward; this makes it possible to spread the negative moment reinforcing bars to make placing of the concrete somewhat easier. Figure 6.18c and d show possibilities for stirrups in beams that occur at the edges of large openings or at the outside edge of the structure. This form of stirrup is used to enhance the torsional resistance of the section and also assists in developing the negative moment resistance in the slab at the edge of the beam.

So-called *closed stirrups*, similar to ties in columns, are sometimes used for T-shaped beams, as shown in Fig. 6.18c and f. These are generally used to improve the torsional resistance of the beam section.

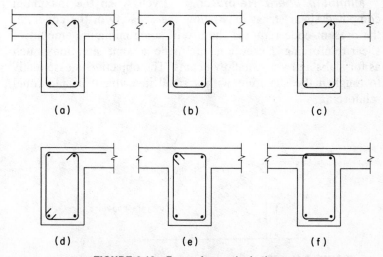

(a)   (b)   (c)

(d)   (e)   (f)

FIGURE 6.18 Forms for vertical stirrups.

Stirrup forms are often modified by designers or by the reinforcing fabricator's detailers to simplify the fabrication and/or the field installation. The stirrups shown in Fig. 6.18d and f are two such modifications of the basic details in Fig. 6.18c and e, respectively.

Beam Shear: General Considerations

The following are some of the general considerations and code requirements that apply to current practices of design for beam shear.

Concrete Capacity.   Whereas the tensile strength of the concrete is ignored in design for flexure, the concrete is assumed to take some portion of the shear in beams. If the capacity of the concrete is not exceeded—as it sometimes is for lightly loaded beams—there may be no need for reinforcing. The typical case, however, is as shown in Fig. 6.19, where the maximum shear V exceeds the capacity of the concrete alone (V_c) and the steel reinforcing is required to absorb the excess, indicated as the shaded portion in the shear diagram.

Minimum Shear Reinforcing.   Even when the maximum computed shear stress falls below the capacity of the concrete, the present code requires the use of some minimum amount of shear reinforcing. Exceptions are made in some situations, such as for slabs and very shallow beams. The objective is essentially to toughen the structure with a small investment in additional reinforcing.

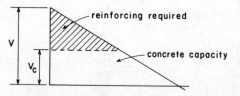

FIGURE 6.19   Sharing of shear resistance in reinforced beams.

Type of Stirrup. The most common stirrups are the simple U shape or closed forms shown in Fig. 6.18, placed in a vertical position at intervals along the beam. It is also possible to place stirrups at an incline (usually 45°), which makes them somewhat more effective in direct resistance to the potential shear cracking near the beam ends (see Fig. 6.15). In large beams with excessively high unit shear stress, both vertical and inclined stirrups are sometimes used at the location of the greatest shear.

Size of Stirrups. For beams of moderate size, the most common size for U-stirrups is a No. 3 bar. These bars can be bent relatively tightly at the corners (small radius of bend) in order to fit within the beam section. For larger beams, a No. 4 bar is sometimes used, its strength (as a function of its cross-sectional area) being almost twice that of a No. 3 bar.

Reinforcing for Narrow Beams. When beams are less than about 10 in. wide, it is not possible to bend a U-shaped stirrup to fit within the beam profile. If shear reinforcing is required, one form that is used is the so-called *ladder* stirrup, shown in Fig. 6.20. This consists of a series of single vertical bars welded to horizontal bars at the top and bottom. A variation on this type of reinforcing consists of using a portion of heavy-gage welded wire fabric.

Spacing of Stirrups. Stirrup spacings are computed (as discussed in the following sections) on the basis of the amount of

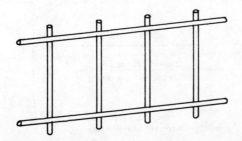

FIGURE 6.20 "Ladder" shear reinforcement in a narrow beam.

reinforcing required for the unit shear stress at the location of the stirrups. A maximum spacing of $d/2$ (i.e., one-half the effective beam depth d) is specified in order to assure that at least one stirrup occurs at the location of any potential diagonal crack (see Fig. 6.15). When shear stress is excessive, the maximum spacing is limited to $d/4$.

Critical Maximum Design Shear. Although the actual maximum shear value occurs at the end of the beam, the code permits the use of the shear stress at a distance of d (effective beam depth) from the beam end as the critical maximum for stirrup design. Thus, as shown in Fig. 6.21, the shear requiring reinforcing is slightly different from that shown in Fig. 6.19.

Total Length for Shear Reinforcing. On the basis of computed shear stresses, reinforcing must be provided along the beam length for the distance defined by the shaded portion of the shear stress diagram shown in Fig. 6.21. For the center portion of the span, the concrete is theoretically capable of the necessary shear resistance without the assistance of reinforcing. However,

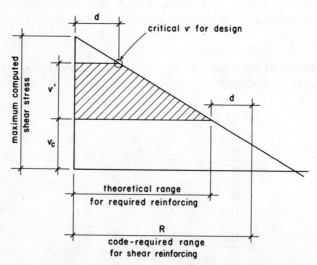

FIGURE 6.21 Layout for shear stress analysis.

the code requires that some reinforcing be provided for a distance beyond this computed cutoff point. The 1963 ACI Code required that stirrups be provided for a distance equal to the effective depth of the beam beyond the cutoff point. The 1989 ACI Code requires that minimum shear reinforcing be provided as long as the computed shear stress exceeds one-half of the capacity of the concrete. However it is established, the total extended range over which reinforcing must be provided is indicated as R on Fig. 6.21.

Design for Beam Shear: Working Stress Method

The following is a description of the procedure for design of shear reinforcing for beams that is in compliance with Appendix A of the 1989 ACI Code (Ref. 1).

Shear stress is computed as

$$v = \frac{V}{bd}$$

where V = total shear force at the section

b = beam width (of the stem for T-shapes)

d = effective depth of the section

For beams of normal weight concrete, subjected only to flexure and shear, shear stress in the concrete is limited to

$$v_c = 1.1 \sqrt{f'_c}$$

When v exceeds the limit for v_c, reinforcing must be provided, complying with the general requirements discussed previously. Although the code does not use the term, we coin the notation of v' for the excess unit shear for which reinforcing is required. Thus

$$v' = v - v_c$$

Required spacing of shear reinforcement is determined as follows. Referring to Fig. 6.22, we note that the capacity in tensile

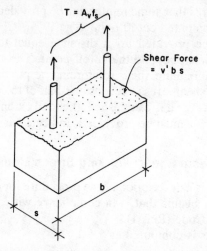

FIGURE 6.22 Spacing consideration for a single stirrup.

resistance of a single, two-legged stirrup is equal to the product of the total steel cross-sectional area times the allowable steel stress. Thus

$$T = A_v f_s$$

This resisting force opposes the development of shear stress on the area s times b, where b is the width of the beam and s is the spacing (half the distance to the next stirrup on each side). Equating the stirrup tension to this force, we obtain the equilibrium equation

$$A_v f_s = b s v'$$

From this equation, we can derive an expression for the required spacing; thus

$$s = \frac{A_v f_s}{v' b}$$

The following example illustrates the procedure for a simple beam.

Example 1. Using the working stress method, design the required shear reinforcing for the simple beam shown in Fig. 6.23. Use $f'_c = 3$ ksi [20.7 MPa] and $f_s = 20$ ksi [138 MPa] and single U-shaped stirrups.

Solution: The maximum value for the shear is 40 kips [178 kN] and the maximum value for shear stress is computed as

$$v = \frac{V}{bd} = \frac{40,000}{12(24)} = 139 \text{ psi } [957 \text{ kPa}]$$

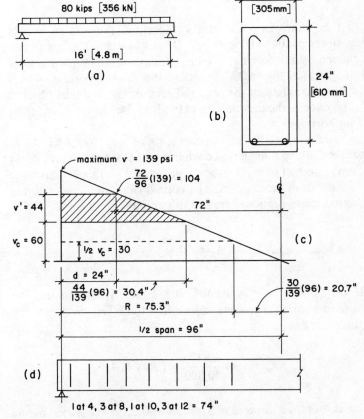

FIGURE 6.23 Stirrup design: Example 1.

We now construct the shear stress diagram for one half of the beam, as shown in Fig. 6.23c. For the shear design, we determine the critical shear stress at 24 in. (the effective depth of the beam) from the support. Using proportionate triangles, this value is

$$\frac{72}{96} (139) = 104 \text{ psi [718 kPa]}$$

The capacity of the concrete without reinforcing is

$$v_c = 1.1 \sqrt{f'_c} = 1.1 \sqrt{3000} = 60 \text{ psi [414 kPa]}$$

At the point of critical stress, therefore, there is an excess shear stress of $104 - 60 = 44$ psi $[718 - 414 = 304$ kPa] that must be carried by reinforcing. We next complete the construction of the diagram in Fig. 6.23c to define the shaded portion, which indicates the extent of the required reinforcing. We thus observe that the excess shear condition extends to 54.4 in [1.382 m] from the support.

In order to satisfy the requirements of the 1989 ACI Code, shear reinforcing must be used wherever the computed unit stress exceeds one-half of v_c. As shown in Fig. 6.23c, this is a distance of 75.3 in. from the support. The code further stipulates that the minimum cross-sectional area of this reinforcing be

$$A_v = 50 \frac{bs}{f_y}$$

If we assume an f_y value of 50 ksi [345 MPa] and use the maximum allowable spacing of one-half the effective depth, the required area is

$$A_v = (50) \frac{12(12)}{50,000} = 0.144 \text{ in.}^2$$

which is less than the area of $2 \times 0.11 = 0.22$ in.^{2} provided by the two legs of the No. 3 stirrup.

For the maximum v' value of 44 ksi, the maximum spacing permitted is determined as

$$s = \frac{A_v f_s}{v'b} = \frac{(0.22 \text{ in.}^2)(20,000 \text{ psi})}{(44 \text{ psi})(12 \text{ in.})} = 8.3 \text{ in.}$$

Since this is less than the maximum allowable of one-half the depth or 12 in., it is best to calculate at least one more spacing at a short distance beyond the critical point. We thus determine that the unit stress at 36 in. from the support is

$$v = \frac{60}{96}(139) = 87 \text{ psi}$$

and the value of v' at this point is $87 - 60 = 27$ psi. The spacing required at this point is thus

$$s = \frac{0.22(20,000)}{27(12)} = 13.6 \text{ in.}$$

which indicates that the required spacing drops to the maximum allowed at less than 12 in. from the critical point. A possible choice for the stirrup spacings is shown in Fig. 6.23d, with a total of eight stirrups that extend over a range of 74 in. from the support. There are thus a total of 16 stirrups in the beam, 8 at each end.

Example 2. Determine the required number and spacings for No. 3 U-stirrups for the beam shown in Fig. 6.24. Use $f'_c = 3$ ksi [20.7 MPa] and $f_s = 20$ ksi [138 MPa].

Solution: As in Example 1, the shear values and corresponding stresses are determined, and the diagram in Fig. 6.24c, is constructed. In this case, the maximum critical shear stress of 89 psi results in a maximum v' value of 29 psi, for which the required spacing is

$$s = \frac{0.22(20,000)}{29(10)} = 15.2 \text{ in.}$$

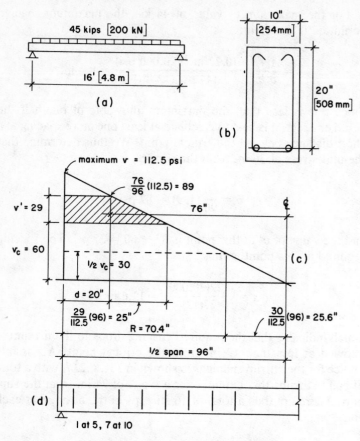

FIGURE 6.24 Stirrup design: Example 2.

Since this value exceeds the maximum limit of $d/2 = 10$ in., the stirrups may all be placed at the limiting spacing, and a possible arrangement is as shown in Fig. 6.24d.

Note that in both Examples 1 and 2 the first stirrup is placed at one-half the required distance from the support.

Example 3. Determine the required number and spacings for No. 3 U-stirrups for the beam shown in Fig. 6.25. Use $f'_c = 3$ ksi [20.7 MPa] and $f_s = 20$ ksi [138 MPa].

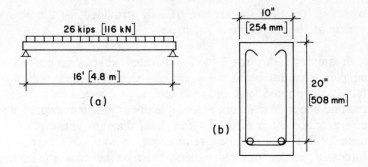

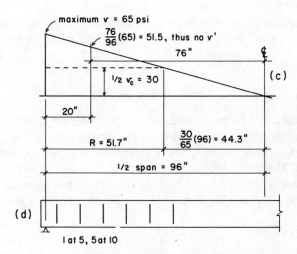

FIGURE 6.25 Stirrup design: Example 3.

Solution: In this case, the maximum critical design shear stress is found to be less than v_c, which in theory indicates that reinforcing is not required. To comply with the code requirement for minimum reinforcing, however, we provide stirrups at the maximum permitted spacing out to the point where the shear stress drops to 30 psi (one-half of v_c). To verify that the No. 3 stirrup is adequate, we compute

$$A_v = (50) \frac{10(10)}{50,000} = 0.10 \text{ in.}^2 \quad \text{(see Example 1)}$$

which is less than the area of 0.22 in. provided, so the No. 3 stirrup at 10-in. spacing is adequate.

Examples 1 through 3 have illustrated what is generally the simplest case for beam shear design—that of a beam with uniformly distributed load and with sections subjected only to flexure and shear. When concentrated loads or unsymmetrical loadings produce other forms for the shear diagram, these must be used for design of the shear reinforcing. In addition, where axial forces of tension or compression exist in the concrete frame, consideration must be given to the combined effects when designing for shear.

When torsional moments exist (twisting moments at right angles to the beam), their effects must be combined with beam shear.

Design for Beam Shear: Strength Method

The requirements and procedures for strength design are essentially similar to those for working stress design. The principal difference is in the use of ultimate resistance as opposed to working stresses at service loads. The basic requirement in strength design is that the modified ultimate resistance of the section be equal to or greater than the factored load. This condition is stated as

$$V_u \leq \phi V_n$$

where V_u = factored shear force at the section
V_n = nominal shear strength of the section

The nominal strength is defined as

$$V_n = V_c + V_s$$

where V_c = nominal strength provided by concrete
V_s = nominal strength provided by reinforcing

The term *nominal strength* is used to differentiate between the computed resistances and the usable value of total resistance, which is reduced for design by the strength reduction factor ϕ.

For members subjected to shear and flexure only, the nominal concrete strength is defined as

$$V_c = 2 \sqrt{f'_c}\, bd$$

Translated into unit stress terms, this means that the limiting nominal shear stress in the concrete is $2 \sqrt{f'_c}$, and when reduced by ϕ, the limiting *working* ultimate strength is $0.85 \times 2 \sqrt{f'_c} = 1.7 \sqrt{f'_c}$.

When shear reinforcing consists of vertical stirrups, the nominal reinforcing strength is defined as

$$V_s = \frac{A_u f_y d}{s}$$

with a limiting value for V_s established as

$$V_s = 8 \sqrt{f'_c}\, bd$$

The following example illustrates the use of strength design methods for shear reinforcing. The problem data are essentially the same as for Example 1, so that a comparison of the design results can be made.

Example 4. Using strength design methods, determine the spacing required for No. 3 U-stirrups for the beam shown in Fig. 6.26. Use $f'_c = 3$ ksi [20.7 MPa] and $f_y = 50$ ksi [345 Mpa].

Solution: The loads shown in Fig. 6.26a are service loads. These must be converted to *factored loads* for strength design, as discussed in Sec. 6.5. We thus determine the factored load to be

$$W_u = 1.4(\text{dead load}) + 1.7(\text{live load})$$

$$= 1.4(40) + 1.7(40)$$

$$= 124 \text{ kips}$$

The maximum shear force is thus 62 kps, and the shear diagram for one-half the beam is as shown in Fig. 6.26c. The critical value for V_u at 24 in. (effective beam depth) from the support is

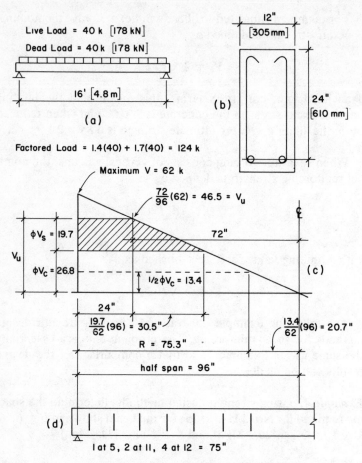

FIGURE 6.26 Stirrup design: Example 4.

determined from proportionate triangles to be 46.5 kips. The usable capacity of the concrete is determined as

$$\phi V_c = \phi 2 \sqrt{f'_c}\, bd$$

$$= 0.85(2 \sqrt{3000})(12)(24)$$

$$= 26,816 \text{ lb or approximately } 26.8 \text{ kips}$$

and for the reinforcing

$$\phi V_s = 46.5 - 26.8 = 19.7 \text{ kips}$$

Therefore,

$$V_s = \frac{19.7}{\phi} = \frac{19.7}{0.85} = 23.18 \text{ kips}$$

and the required spacing is determined from

$$V_s = \frac{A_v f_y d}{s}$$
$$s = \frac{A_v f_y d}{V_s} = \frac{0.22(50)(24)}{23.18} = 11.4 \text{ in.}$$

Referring to Example 1, we may see that this value is larger than that computed by the working stress method; thus the strength design is somewhat less conservative for this example. A possible choice of stirrup spacings is that shown in Fig. 6.26d, using seven stirrups at each end of the beam.

To verify that the value for V_s is within the limit previously given, we compute the maximum value of

$$V_s = 8 \sqrt{f_c'} \, bd = 8 \sqrt{3000} \, (12)(24) = 126 \text{ kips}$$

which is far from critical.

PROBLEM 6.9.A. A concrete beam similar to that shown in Fig. 6.23 sustains a total load of 60 kips [267 kN] on a span of 24 ft. [7.32 m]. Determine the layout for a set of No. 3 U-stirrups using $f_s = 20$ ksi [138 MPa] and $f_c' = 3000$ psi [20.7 MPa]. The section dimensions are $b = 12$ in. [305 mm] and $d = 26$ in [660 mm].

PROBLEM 6.9.B. Determine the layout for a set of No. 3 U-stirrups for a beam with the same data as Problem 6.9.A, except the total load on the beam is 30 kips [133 kN].

6.10 DEVELOPMENT OF REINFORCEMENT

Bond stresses are developed on the surfaces of reinforcing bars whenever some structural action requires the steel and concrete to interact. In times past, working stress procedures included the establishment of allowable stresses for bond and the computation of bond stresses for various situations. At present, however, the codes deal with this problem as one of development length. This chapter presents a discussion of bond stress situations and the current practices in establishing required lengths for the development of reinforcement.

Bonding

The basic concept of bond stress development is illustrated by the example shown in Fig. 6.27, in which a steel bar is embedded in a block of concrete and is required to resist a pull-out tension force. Figure 6.27b shows the static equilibrium relationship for the steel bar, with the pull-out force developed as the product of a tensile stress times the area of the bar cross section $[f_s(\pi D^2/4)]$ and the resisting force developed by a bond stress (u) operating on the surface of the bar $[u(\pi D)L]$. By equating these two forces, we can derive an expression either for the unit bond stress or the required embedment length for a limiting bond stress.

Bond stress development is affected by a number of considerations; some of the major ones are the following:

Grade of Steel.   As the f_y of the steel is increased, the allowable f_s value will also increase, requiring the development of higher bond stresses or the need for greater embedment lengths.

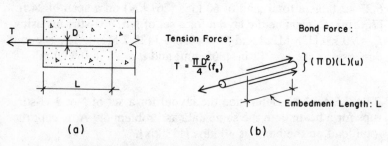

$$T = \frac{\pi D^2}{4}(f_s)$$

(a)   (b)

FIGURE 6.27 Development of bonding stress.

Strength of Concrete.   In general, as f'_c is increased, the capability for development of bond stress is also increased.

Bar Size.   Consideration of the expression for the tension force in the bar in Fig. 6.27 will indicate that the force capability of the bar increases with the square of the diameter. On the other hand, the resistance developed by bond stress increases only linearly with increase of the bar diameter. Thus bond stresses tend to be more critical on bars of large diameter.

Concrete Encasement.   The bonding force must be developed in the concrete mass around each bar. This development is limited when this mass is constrained due to closely spaced groups of bars or where bars are placed close to the edge of the concrete member.

Location of Bars.   When concrete is poured into forms and cured into hardened state, the concrete near the bottom of the member tends to develop slightly higher quality than that near the top. The weight of the concrete mass above produces a denser material in the lower concrete, and the exposed top surface tends to dry more rapidly, resulting in less well-cured concrete near the top. This difference in quality affects the potential for bond resistance, so some adjustment is made for bars placed near the top (such as reinforcement for negative moment in beams).

Development of Reinforcement

The ACI Code defines *development length* as the length of embedment required to develop the design strength of the reinforcing at a critical section. For beams, critical sections occur at points of maximum stress and at points within the span where some of the reinforcement terminates or is bent up or down. For a uniformly loaded simple span beam, one critical section is at midspan, where the bending moment is a maximum. The tensile reinforcing required for flexure at this point must extend on both sides a sufficient distance to develop the stress in the bars; however, except for very short spans with large bars, the bar lengths will ordinarily be more than sufficient.

In the simple beam, the bottom reinforcing required for the maximum moment at midspan is not entirely required as the moment decreases toward the end of the span. It is thus sometimes the practice to make only part of the midspan reinforcing continuous for the whole beam length. In this case it may be necessary to assure that the bars that are of partial length are extended sufficiently from the midspan point and that the bars remaining beyond the cutoff point can develop the stress required at that point.

When beams are continuous through the supports, top reinforcing is required for the negative moments at the supports. These top bars must be investigated for the development lengths in terms of the distance they extend from the supports.

For tension reinforcing consisting of bars of No. 11 size and smaller, the code specifies a minimum length for development (l_d) as follows:

$$l_d = 0.04 \, A_b \, \frac{f_y}{\sqrt{f'_c}}$$

but not less than $0.0004d_b f_y$ or 12 in. In these formulas A_b is the cross-sectional area of the bar and d_b is the bar diameter.

Modification factors for l_d are given for various situations, as follows:

For top bars in horizontal members with at least
    12 in. of concrete below the bars   1.3

For flexural reinforcement that is in excess of
    that required by computations $\dfrac{A_s \text{ required}}{A_s \text{ provided}}$

Additional modification factors are given for light-weight concrete, for bars encased in spirals, and for bars with f_y in excess of 60 ksi.

Table 6.6 gives values for minimum development lengths for tensile reinforcing, based on the requirements of the ACI Code. The values listed under "other bars" are the unmodified length requirements; those listed under "top bars" are increased by the

TABLE 6.6 Minimum Development Length for Tensile Reinforcement (in.)[a]

| | f_y = 40 ksi [276 MPa] | | | | f_y = 60 ksi [414 MPa] | | | |
| | f'_c = 3 ksi [20.7 MPa] | | f'_c = 4 ksi [27.6 MPa] | | f'_c = 3 ksi [20.7 MPa] | | f'_c = 4 ksi [27.6 MPa] | |
Bar Size	Top Bars[b]	Other Bars	Top Bars[b]	Other Bars	Top Bars[b]	Other Bars	Top Bars[b]	Other Bars
3	12	12	12	12	12	12	12	12
4	12	12	12	12	16	12	16	12
5	13	12	12	12	20	15	20	15
6	17	13	15	12	25	19	23	18
7	23	18	20	15	34	26	30	23
8	30	23	26	20	45	35	39	30
9	38	29	33	25	57	44	50	38
10	48	37	42	32	73	56	63	48
11	60	46	51	40	89	68	77	59

[a] Lengths are based on requirements of the ACI Code (Ref. 1).
[b] Horizontal bars so placed that more than 12 in. [305 mm] of concrete is cast in the member below the reinforcement.

modification factor for this situation. Values are given for two concrete strengths and for the two most commonly used grades of tensile reinforcing.

Hooks

When details of the construction restrict the ability to extend bars sufficiently to produce required development lengths, partial development can sometimes by achieved by use of a hooked end. Section 12.5 of the ACI Code provides a means by which a so-called standard hook may be evaluated in terms of an equivalent development length. Bending requirements for standard hooks are given in Chapter 7 of the ACI Code. Bar ends may be bent at 90, 135, or 180° to produce a hook. The 135° bend is used only for ties and stirrups, which normally consist of relatively small diameter bars (see Fig. 4.2).

Table 6.7 gives values for standard hooks, using the same variables for f'_c and f_y that are used in Table 6.6. The table values given are in terms of the equivalent development length provided by the hook. Comparison of the values in Table 6.7 with those given for the unmodified lengths ("Other") in Table 6.6 will show that the hooks are mostly capable of only partial development. The development length provided by a hook may be added to whatever development length is provided by extension of the bar, so that the total development may provide for full utilization of the bar tension capacity (at f_y) in many cases. The following example illustrates the use of the data from Tables 6.6 and 6.7 for a simple situation.

Example 1. The negative moment in the short cantilever shown in Fig. 6.28 is resisted by the steel bar in the top of the beam. Determine whether the development of the reinforcing is adequate. Use $f'_c = 3$ ksi [20 MPa] and $f_y = 60$ ksi [414 MPa].

Solution: The maximum moment in the cantilever is produced at the face of the support; thus the full tensile capacity of the bar should be developed on both sides of this section. In the beam

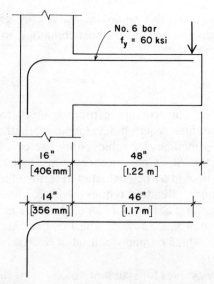

FIGURE 6.28 Analysis for bar development: Example 1.

itself the condition is assumed to be that of a "top bar," for which Table 6.6 yields a required minimum development length of 27 in., indicating that the length of 46 in. provided is more than adequate. Within the support, the condition is unmodified, and the requirement is for a length of 19 in. The actual extended development length provided within the support is 14 in., which is measured as the distance to the end of the hooked bar end, as shown in the figure. If the hooked end qualifies as a *standard hook* (in accordance with the requirements of Chapter 7 of the ACI Code), the equivalent development length provided (from Table 6.7) is 9.5 in. Thus the total development provided by the combination of extension and hooking is 14 + 9.5 = 23.5 in., which exceeds the requirement of 19 in., so the development is adequate.

In a real situation, it is probably not necessary to achieve the full development lengths given in Table 6.6, since bar selection often results in some slight excess in the actual steel cross-sectional area provided. In such a case, the required development length can be reduced by the modification factor.

Development of Compressive Reinforcement

The discussion of development length so far has dealt with tension bars only. Development length in compression is, of course, a factor in column design and in the design of beams reinforced for compression.

The absence of flexural tension cracks in the portions of beams where compression reinforcement is employed, plus the beneficial effect of the end bearing of the bars on the concrete, permit shorter developmental lengths in comparison than in tension. The ACI Code prescribes that l_d for bars in compression shall be computed by the formula

$$l_d = \frac{0.02 f_y d_b}{\sqrt{f'_c}}$$

but shall not be less than $0.0003 f_y d_b$ or 8 in., whichever is greater. Table 6.8 lists compression bar development lengths for a few combinations of specification data.

TABLE 6.7 Equivalent Embedment Lengths of Standard Hooks (in.)

Bar Size	$f_y = 40$ ksi [276 MPa]		$f_y = 60$ ksi [414 MPa]	
	$f'_c = 3$ ksi [20.7 MPa]	$f'_c = 4$ ksi [27.6 MPa]	$f'_c = 3$ ksi [20.7 MPa]	$f'_c = 4$ ksi [27.6 MPa]
3	3.0	3.4	4.4	5.1
4	3.9	4.5	5.9	6.8
5	4.9	5.7	7.4	8.5
6	6.3	6.8	9.5	10.2
7	8.6	8.6	12.9	12.9
8	11.4	11.4	17.1	17.1
9	14.4	14.4	21.6	21.6
10	18.3	18.3	24.4	24.4
11	22.5	22.5	26.2	26.2

TABLE 6.8 Minimum Development Length for Compressive Reinforcement (in.)

Bar Size	$f_y = 40$ ksi [276 MPa]		$f_y = 60$ ksi [414 MPa]	
	$f'_c = 3$ ksi [20.7 MPa]	$f'_c = 4$ ksi [27.6 MPa]	$f'_c = 3$ ksi [20.7 MPa]	$f'_c = 4$ ksi [27.6 MPa]
3	8.0	8.0	8.0	8.0
4	8.0	8.0	11.0	9.5
5	9.2	8.0	13.7	11.9
6	10.9	9.5	16.4	14.2
7	12.8	11.1	19.2	16.6
8	14.6	12.7	21.9	19.0
9	16.5	14.3	24.8	21.5
10	18.5	16.1	27.8	24.1
11	20.6	17.9	31.0	26.8
14			37.1	32.1
18			49.5	42.8

In reinforced concrete columns both the concrete and the steel bars share the compression force. Ordinary construction practices require the consideration of various situations for development of the stress in the reinforcing bars. Figure 6.29 shows a multistory concrete column with its base supported on a concrete footing. With reference to the illustration, we note the following.

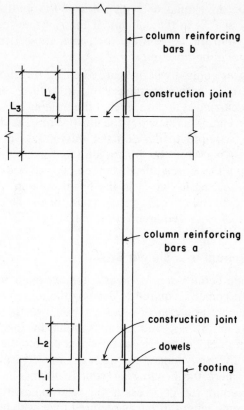

FIGURE 6.29 Bar development considerations for columns.

1. The concrete construction is ordinarily produced in multiple, separate pours, with construction joints between the separate pours occurring as shown in the illustration.

2. In the lower column, the load from the concrete is transferred to the footing in direct compressive bearing at the joint between the column and footing. The load from the reinforcing must be developed by extension of the reinforcing into the footing: distance L_1 in the illustration. Although it may be possible to place the column bars in position during pouring of the footing to achieve this, the common practice is to use dowels, as shown in the illustration. These dowels

must be developed on both sides of the joint: L_1 in the footing and L_2 in the column. If the f'_c value for both the footing and the column are the same, these two required lengths will be the same.

3. The lower column will ordinarily be cast together with the supported concrete framing above it, with a construction joint occurring at the top level of the framing (bottom of the upper column), as shown in the illustration. The distance L_3 is that required to develop the reinforcing in the lower column—bars a in the illustration. As for the condition at the top of the footing, the distance L_4 is required to develop the reinforcing in bars b in the upper column. L_4 is more likely to be the critical consideration for the determination of the extension required for bars a.

Bar Development in a Simple Beam

The ACI Code defines *development length* as the length of embedded reinforcement required to develop the design strength of the reinforcement at a critical section. Critical sections occur at points of maximum stress and at points within the span at which adjacent reinforcement terminates or is bent up into the top of the beam. For a uniformly loaded simple beam, one critical section is at midspan where the bending moment is maximum. This is a point of maximum tensile stress in the reinforcement (peak bar stress) and some length of bar is required over which the stress can be developed. Other critical sections occur between midspan and the reactions at points where some bars are cut off because they are no longer needed to resist the bending moment; such terminations create peak stress in the remaining bars that extend the full length of the beam.

Example 2. A uniformly loaded rectangular beam is used on a simple span of 16 ft. It sustains a theoretical bending moment of $M_t = 1944$ kip-in., induced by service dead and live loads of 29 and 19 kips, respectively; these produce a factored uniform load of $W_u = 73.0$ kips. The beam is 10 in. wide and has an effective depth of 14 in. The tension reinforcement consists of four No. 8 bars, all of which extend the entire length of the beam. Compute the required development length of the reinforcement if $f_y = 60,000$ psi and $f_c' = 4000$ psi.

Solution: Referring to Table 2.2, the area of an individual No. 8 bar is found to be 0.79 in.^{2}. Then

$$l_d = \frac{0.04 A_b f_y}{\sqrt{f_c'}} = \frac{0.04(0.79)(60,000)}{\sqrt{4000}} = 30 \text{ in.}$$

Because all bars extend the full length of the beam, an embedment length (from section of peak bar stress at midspan) equal to 8 ft., or 96 in., is provided. This greatly exceeds the required development length of 30 in., and consequently, l_d is not a critical factor in this beam.

Example 3. If two of the four bars in the beam of Example 1 are cut off short of each end at sections beyond which they are no longer needed for bending moment, compute the required development length of the other two bars that continue into the supports.

Solution: (1) The parabolic curve in Fig. 6.30 represents the bending moment diagram for this uniformly loaded simple beam.

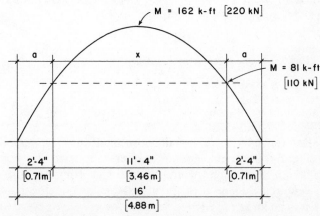

FIGURE 6.30 Bar development in a simple beam.

Although presumably there is the usual small discrepancy between A_s required and that actually supplied (due to use of standard bar sizes), we neglect this difference and assume that the potential resisting moment of the four bars just matches the 1944 kip-in. developed by the loading.

(2) The dashed line in Fig. 6.30 indicates the resisting moment of half the total reinforcement. Where this line intersects the bending moment curve two of the four bars are no longer necessary and may be cut off. For a uniformly loaded simple beam this intersection falls between $\frac{1}{6}$ and $\frac{1}{7}$ of the span length from the support. Taking $L/7$ as the approximate position, the distance in this case is $16 \div 7 = 2.29$ ft or 2 ft 4 in. (28 in.).

(3) At this cutoff point the peak stress in the *continuing* bars is the same as at midspan, and consequently a development length of 30 in. (see Example 2 or Table 6.6) is required beyond this critical section. Because the section occurs 28 in. from the end of the span, l_d is critical by this test and a standard hook must be made at the end of the bars. Referring to Table 6.7, we note that the equivalent embedment length of a standard hook on a bottom No. 8 bar for our specification data is 17.1 in. This is more than adequate to cover the $30 - 28 = 2$ in. of extra length required here.

(4) An additional condition must be satisfied at simple supports. The diameter of the reinforcement must be small enough so that the computed development length of the bar will not exceed $(M_t/V_u) + l_a$. This may be expressed as

$$l_d \lessgtr \frac{M_t}{V_u} + l_a$$

The standard hook must be retained, however, to satisfy the other requirements of step 3.

(5) Reference to Fig. 6.30 will show that if the *terminated* bars were actually cut off at the points of intersection with the moment curve their overall length would be 11 ft 4 in. However, the ACI Code requires that reinforcement extend beyond the point at which it is no longer needed in bending for a distance equal to the effective depth of the beam or 12 bar diameters, whichever is

greater. In our case the effective depth is 14 in., and 12 times the nominal diameter of a No. 8 bar is 12 in. Therefore, the actual length of the terminated bars will be 11 ft 4 in., plus twice 14 in., or 13 ft 8 in.

In practice it would have to be decided whether the cost of fabricating hooks on both ends of the two continuing bars might exceed that of the extra steel used if all four bars continued into the supports (as in Example 2). The simpler fabrication involved and the greater ease of placing the reinforcement in the forms would probably favor selection in this case of four full-length straight bars.

Bar Development in Continuous Beams

When beams are continuous through their supports, the negative moments at the supports will require that bars be placed in the top of the beams. Within the span, bars will be required in the bottom of the beam for the positive moments. While the positive moment will go to zero at some distance from the supports, the codes require that some of the positive moment reinforcing be extended for the full length of the span and a short distance into the support.

Figure 6.31 shows a possible layout for reinforcing in a beam with continuous spans and a cantilevered end at the first support. Referring to the notation in the illustration, we make the following observations.

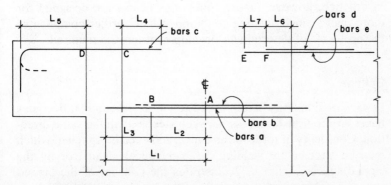

FIGURE 6.31 Development lengths in continuous beams.

1. Bars a and b are provided for the maximum moment of positive sign that occurs somewhere near the beam midspan. If all these bars are made full length (as shown for bars a), the length L_1 must be sufficient for development (this situation is seldom critical). If bars b are partial length, as shown in the illustration, then length L_2 must be sufficient to develop bars b and length L_3 must be sufficient to develop bars a. As was discussed for the simple beam, the partial length bars must actually extend beyond the theoretical cutoff point (B in the illustration) and the true length must include the dashed portions indicated for bars b.

2. For the bars at the cantilevered end, the distances L_4 and L_5 must be sufficient for development of bars c. L_4 is required to extend beyond the actual cutoff point of the negative moment by the extra length described for the partial length bottom bars. If L_5 is not adequate, the bar ends may be bent into the 90° hook as shown or the 180° hook shown by the dashed line.

3. If the combination of bars shown in the illustration is used at the interior support, L_6 must be adequate for the development of bars d and L_7 adequate for the development of bars e.

For a single loading condition on a continuous beam it is possible to determine specific values of moment and their location along the span, including the locations of points of zero moment. In practice, however, most continuous beams are designed for more than a single loading condition, which further complicates the problems of determining development lengths required.

Splices

In various situations in reinforced concrete structures it becomes necessary to transfer stress between steel bars in the same direction. Continuity of force in the bars is achieved by splicing, which may be affected by welding, by mechanical means, or by the lapped splice. Figure 6.32 illustrates the concept of the lapped splice, which consists essentially of the development of both bars

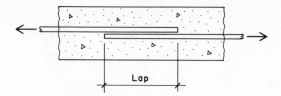

FIGURE 6.32 The lapped splice.

within the concrete. The length of the lap becomes the development length for both bars. Because a lapped splice is usually made with the two bars in contact, the lapped length must usually be somewhat greater than the simple development length required in Table 6.6.

Sections 12.14 to 12.19 of the ACI Code give requirements for various types of splices. For a simpler tension lap splice, the full development of the bars requires a lap length of 1.7 times that required for simple development of the bars. Lap splices are generally limited to bars of No. 11 size and smaller.

For pure tension members, lapped splicing is not permitted, and splicing must be achieved by welding the bars or by some mechanical connection. End-to-end butt welding of bars is usually limited to compression splicing of large diameter bars with high f_y for which lapping is not feasible.

When members have several reinforcing bars that must be spliced, the splicing must be staggered. In general, splicing is not desirable, and is to be avoided where possible. Because bars are obtainable only in limited lengths, however, some situations unavoidably involve splicing. Horizontal reinforcing in long walls is one such case.

Ties for Hooked Bars

Where hooks are required for bar development at the discontinuous end of a beam, it may be necessary to provide ties or closed stirrups to prevent splitting of the concrete in the vicinity of the hooks. This requirement is defined in Sec. 12.5.4 of the 1989 ACI Code. Situations of this type may occur at the ends of simple

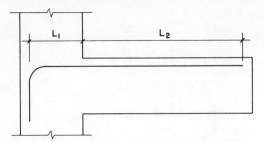

FIGURE 6.33 Reference figure for Problem 6.10.

beams or extended beam ends as shown at the left end of the beam in Fig. 6.30.

Developed Anchorage for Frame Continuity

In concrete rigid frame structures the engagement of the frame members at joints between columns and beams sometimes requires special attention in the detailing of reinforcement. A particular concern is that of the potential for the beams to pull loose from the columns, an action typically resisted by the extended reinforcing bars from the beam ends. In addition to the usual concerns for bar development, some special detailing to enhance the anchoring of the bars may be indicated. This is a matter of particular note in resistance to seismic effects, and is discussed in the development of the framed structures in Chapter 12.

PROBLEM 6.10 A,B,C,D. Determine whether the bar shown in Fig. 6.33 is adequately anchored for development of the full f_y capacity of the steel.

	f'_c		f_y		L_2		L_1		Bar
	ksi	MPa	ksi	MPa	in.	mm	in.	mm	Size
A	3	20.7	60	414	36	914	12	305	8
B	4	27.6	60	414	42	1067	16	206	10
C	3	20.7	40	276	38	965	18	457	9
D	4	27.6	40	276	35	889	17	432	9

7

FLAT-SPANNING SYSTEMS

There are many different concrete systems that can be used to achieve flat spans. These are used most often for floor structures, which typically require a dead flat form. However, in buildings with an all concrete structure, they may also be used for roofs. Sitecast systems generally consist of one of the following basic types:

1. One-way solid slab and beam.
2. Two-way solid slab and beam.
3. One-way joist construction.
4. Two-way flat slab or flat plate without beams.
5. Two-way joist construction, called waffle construction.

Each system has its own distinct advantages and limits and some range of logical use, depending on required spans, general layout of supports, magnitude of loads, required fire ratings, and cost limits for design and construction.

The floor plan of a building and its intended usage determine loading conditions and the layout of supports. Also of concern are requirements for openings for stairs, elevators, large ducts, skylights, and so on, as these result in discontinuities in the otherwise commonly continuous systems. Whenever possible, columns and bearing walls should be aligned in rows and spaced at regular intervals in order to simplify design and construction and lower costs. However, the fluid concrete can be molded in forms not possible for wood or steel, and many very innovative, sculptural systems have been developed as takeoffs on these basic ones.

7.1 SLAB AND BEAM SYSTEMS

The most widely used and most adaptable cast-in-place concrete floor system is that which utilizes one-way solid slabs supported by one-way spanning beams. This system may be used for single spans, but occurs more frequently with multiple-span slabs and beams in a system such as that shown in Fig. 7.1. In the example shown, the continuous slabs are supported by a series of beams that are spaced at 10 ft center to center. The beams, in turn, are supported by a girder and column system with columns at 30-ft centers, every third beam being supported directly by the columns and the remaining beams being supported by the girders.

Because of the regularity and symmetry of the system shown in Fig. 7.1, there are relatively few different elements in the system, each being repeated several times. While special members must be designed for conditions that occur at the outside edge of the system and at the location of any openings for stairs, elevators, and so on, the general interior portions of the structure may be determined by designing only six basic elements: S1, S2, B1, B2, G1, and G2, as shown in the framing plan.

In computations for reinforced concrete, the span length of freely supported beams (simple beams) is generally taken as the distance between centers of supports or bearing areas; it should not exceed the clear span plus the depth of beam or slab. The span length for continuous or restrained beams is taken as the

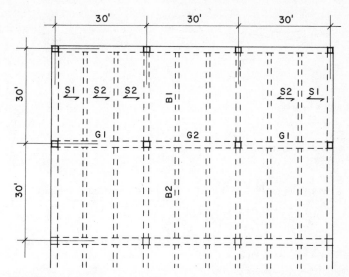

FIGURE 7.1 Framing layout plan for a typical slab-and-beam system.

clear distance between faces of supports. For a simple beam, that is, a single span having no restraint at the supports, the maximum bending moment for a uniformly distributed load is at the center of the span, and its magnitude is $M = WL/8$. The moment is zero at the supports and is positive over the entire span length. In continuous beams, however, negative bending moments are developed at the supports and positive moments at or near midspan. This may be readily observed from the exaggerated deformation curve of Fig. 7.2a. The exact values of the bending moments depend on several factors, but in the case of approximately equal spans supporting uniform loads, when the live load does not exceed three times the dead load, the bending moment values given in Fig. 7.2 may be used for design.

The values given in Fig. 7.2 are in general agreement with the recommendations of Chapter 8 of the ACI Code. These values have been adjusted to account for partial live loading of multiple-span beams. Note that these values apply only to uniformly loaded beams. Chapter 8 of the ACI Code also gives some factors

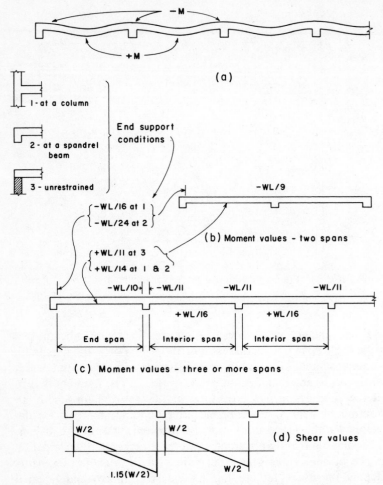

FIGURE 7.2 Approximate design factors for continuous beams.

for end-support conditions other than the simple supports shown in Fig. 7.2.

Design moments for continuous-span slabs are given in Fig. 7.3. Where beams are relatively large and the slab spans are small, the rotational (torsional) stiffness of the beam tends to minimize the effect of individual slab spans on the bending in adjacent spans. Thus most slab spans in the slab-and-beam systems tend to function much like individual spans with fixed ends.

7.2 DESIGN OF A ONE-WAY CONTINUOUS SLAB

The general design procedure for a one-way solid slab was illustrated in Sec. 6.8. The example given there is for a simple span slab. The following example illustrates the procedure for the design of a continuous solid one-way slab.

Example. A solid one-way slab is to be used for a framing system similar to that shown in Fig. 7.1. Column spacing is 30 ft, with evenly spaced beams occurring at 10 ft center to center. Superimposed loads on the structure (floor live load plus other construction dead load) are a total of 138 psf. Use $f'_c = 3$ ksi [20.7 MPa] and grade 60 reinforcement with $f_y = 60$ ksi [414 MPa] and $f_s = 24$ ksi [165 MPa]. Determine the thickness for the slab and pick its reinforcement.

Solution: To find the slab thickness, we consider three factors: the minimum thickness for deflection, the minimum effective depth for the maximum moment, and the minimum effective depth for the maximum shear. For all of these we must first determine the span of the slab. For design purposes this is taken as the clear span, which is the dimension from face to face of the supporting beams. With the beams at 10-ft centers, this dimension is 10 ft, less the width of one beam. Since the beams are not given, we will assume a dimension for them. In practice we would proceed from the slab design to the beam design, after which the assumed dimension could be verified. For this example we will assume a beam width of 12 in., yielding a clear span of 9 ft.

We consider first the minimum thickness required for deflection. If the slabs in all spans have the same thickness (which is the most common practice), the critical slab is the end span, since there is no continuity of the slab beyond the end beam. While the beam will offer some restraint, it is best to consider this as a simple support; thus we use the factor of $L/30$ from Table 7.2:

$$\text{minimum } t = \frac{L}{30} = \frac{9 \times 12}{30} = 3.6 \text{ in.}$$

We will assume here that fire-resistive requirements make it desirable to have a relatively thick slab and so will choose a 5-in. overall thickness, for which the dead weight of the slab will be

$$w = \frac{5}{12} \times 150 = 62 \text{ psf}$$

and the total design load is thus $62 + 138 = 200$ psf.

We next consider the maximum bending moment. Inspection of the moment values given in Fig. 7.3 shows the maximum value to be $\frac{1}{10} wL^2$. With the clear span and the loading as determined, the maximum moment is thus

$$M = \tfrac{1}{10} wL^2 = (0.1)(200)(9)^2 = 1620 \text{ ft-lb}$$

This moment value should now be compared to the balanced capacity of the design section, using the relationships discussed for rectangular beams in Sec. 6.4. For this computation we must assume an effective depth for the design section. This dimension will be the slab thickness minus the concrete cover and one-half the bar diameter. With the bars not yet determined, we will assume an approximate effective depth of the slab thickness minus 1.0 in.; this will be exactly true with the usual minimum cover of $\frac{3}{4}$

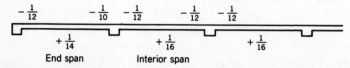

FIGURE 7.3 Approximate design factors for continuous solid slabs with spans of 10 ft or less.

in. and No. 4 bar. Then using the balanced moment R factor from Table 6.1, the maximum resisting moment for the 12-in.-wide design section is

$$M_R = Rbd^2 = (0.204)(12)(4)^2 = 39.17 \text{ kip-in.}$$

or

$$M_R = 39.17 \times \frac{1000}{12} = 3264 \text{ ft-lb}$$

As this value is in excess of the required maximum moment, the slab will be adequate for concrete flexural stress.

It is not practicable to use shear reinforcement in one-way slabs, and consequently the maximum unit shear stress must be kept within the limit for the concrete alone. The usual procedure is to check the shear stress with the effective depth determined for bending before proceeding to find A_s. Except for very short span slabs with excessively heavy loadings, shear stress is seldom critical.

For an interior span, the maximum shear will be $wL/2$, but for the end span it is the usual practice to consider some unbalanced condition for the shear due to the discontinuous end. We therefore use a maximum shear of $1.15wL/2$, or an increase of 15% over the simple beam shear value. Thus

$$\text{maximum shear} = V = 1.15wL/2$$
$$= 1.15(200)(9/2) = 1035 \text{ lb}$$

and

$$\text{maximum shear stress} = v = \frac{V}{bd} = \frac{1035}{12 \times 4} = 22 \text{ psi}$$

This is considerably less than the limit for the concrete alone ($v_c = 1.1 \sqrt{f'_c} = 60$ psi), so the assumed slab thickness is not critical for shear stress.

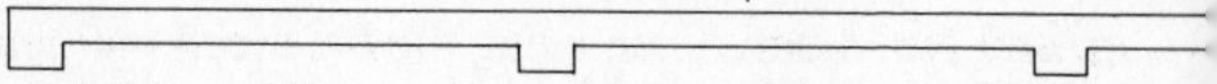

Moment Coefficient: C =	−1/12	+1/14	−1/10	−1/12	+1/16	−1/12	−1/12
Required A$_s$/ft (in.)^2 A$_s$ = 2.25 C	0.1875	0.161	0.225		0.141	0.1875	
Required spacing of reinforcing: (in.) ^{a}							
with No. 3 bars —	7	8 1/2	6		9 1/2	7	
No. 4 —	12	14	10		17	12	
No. 5 —	19	22	16		22	19	

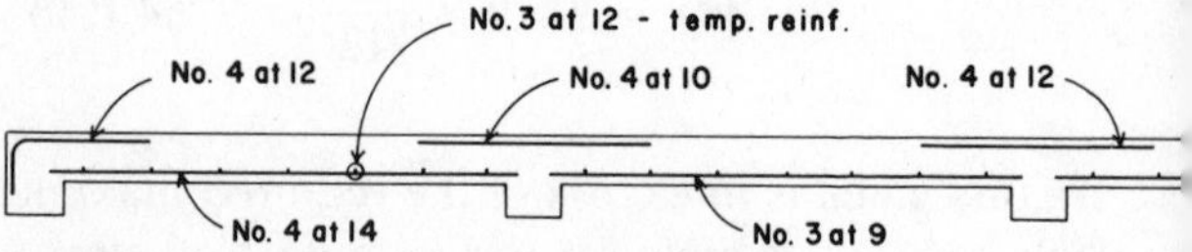

^{a} Maximum of 3t = 15 in.

FIGURE 7.4 Design of the continuous slab.

Having thus verified our choice for the slab thickness, we may now proceed with the design of the reinforcing. For a balanced section, Table 6.1 yields a value of 0.886 for the j factor. However, since all of our reinforced sections will be classified as under-reinforced (actual moment less than the balanced limit), we will use a slightly higher value, say 0.90, for j in the design of the reinforcing.

Referring to Fig. 7.4, we note that there are five critical locations for which a moment must be determined and the required steel area computed. Reinforcing in the top of the slab must be computed for the negative moments at the end support, at the first interior beam, and at the typical interior beam. Reinforcing in the bottom of the slab must be computed for the positive moments at midspan in the first span and the typical interior spans. The design for these conditions is summarized in Fig. 7.4. For the data displayed in the figure we note the following:

maximum spacing of reinforcing = $3 \times t = 3 \times 5 = 15$ in.

maximum moment = M = (moment factor C)(wL^2)

$$= C(200)(9)^2 \times 12$$

$$= C(194,400) \qquad \text{(in in.-lb units)}$$

$$\text{required } A_s = \frac{M}{f_s jd} = \frac{C(194,400)}{(24,000)(0.9)(4)} = 2.25C$$

Using data from Table 6.4, Fig. 7.4 shows required spacings for No. 3, 4, and 5 bars. A possible choice for the slab reinforcement, using all straight bars, is shown at the bottom of the figure.

PROBLEM 7.2.A. A solid one-way slab is to be used for a framing system similar to that shown in Fig. 7.1. Column spacing is 36 ft [11 m], with regularly spaced beams occurring at 12 ft [3.66 m] center to center. Superimposed load on the structure is a total of 180 psf [8.62 kN/m²]. Use $f'_c = 3$ ksi [20.7 MPa] and grade 40 reinforcing with $f_y = 40$ ksi [276 MPa] and $f_s = 20$ ksi [138 MPa]. Determine the thickness for the slab and select the size and spacing for the bars.

7.3 BEAM DESIGN: GENERAL CONSIDERATIONS

The design of a single beam involves a large number of pieces of data, most of which are established for the system as a whole, rather than individually for each beam. System-wide decisions usually include those for the type of concrete and its design strength (f'_c), the type of reinforcing steel (f_y), the cover required for the necessary fire rating, and various generally used details of forming and reinforcement. Most beams occur in conjunction with solid slabs that are poured monolithically with the beams. Slab thickness is established by the structural requirements of the spanning action between beams and by various concerns, such as those for fire rating, thermal and acoustic separation, type of reinforcement, and so on. Design of a single beam is usually limited to determination of the following:

1. Choice of shape and dimensions of the beam cross section.
2. Selection of the type, size, and spacing of shear reinforcement.

3. Selection of the flexural reinforcement to satisfy requirements based on the variation of moment along the beam span.

The following are some factors that must be considered in effecting these decisions.

Beam Shape

Figure 7.5 shows the most common shapes used for beams in poured-in-place construction. The single, simple rectangular section is actually uncommon, but does occur in some situations. Design of the concrete section consists of selecting the two dimensions: the width b and the overall height or depth h.

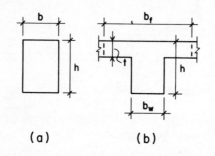

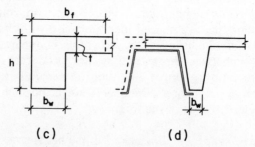

FIGURE 7.5 Common shapes for beams.

As mentioned previously, beams occur most often in conjunction with monolithic slabs, resulting in the typical T shape shown in Fig. 7.5b or the L shape shown in Fig. 7.5c. The full T shape occurs at the interior portions of the system, while the L shape occurs at the outside edge of the system or at the side of large openings. As shown in the illustration, there are four basic dimensions for the T and L that must be established in order to fully define the beam section.

t is the slab thickness; it is ordinarily established on its own, rather than as a part of the single beam design.

h is the overall beam stem depth, corresponding to the same dimension for the rectangular section.

b_w is the beam stem width, which is critical for consideration of shear and for problems of fitting reinforcing into the section.

b_f is the so-called *effective width* of the flange, which is the portion of the slab assumed to work with the beam.

A special beam shape is that shown in Fig. 7.5d. This occurs in concrete joist and waffle construction when "pans" of steel or reinforced plastic are used to form the concrete, the taper of the beam stem being required for easy removal of the forms. The smallest width dimension of the beam stem is ordinarily used for the beam design in this situation.

Beam Width

The width of a beam will affect its resistance to bending. Consideration of the flexure formulas given in Secs. 6.4 and 6.5 will show that the width dimension affects the bending resistance in a linear relationship (double the width and you double the resisting moment, etc.). On the other hand, the resisting moment is affected by the *square* of the effective beam depth. Thus efficiency—in terms of beam weight or concrete volume—will be obtained by striving for deep, narrow beams, instead of shallow, wide ones. (Just as a 2 × 8 joist is more efficient than a 4 × 4 joist in wood.)

Beam width also relates to various other factors, however, and these are often critical in establishing the minimum width for a given beam. The formula for shear stress ($v = V/bd$) indicates that the depth is less effective in shear resistance than in moment resistance. Placement of reinforcing bars is also a problem in narrow beams. Table 7.1 gives minimum beam widths required for various bar combinations, based on considerations of bar spacing (Sec. 4.4), minimum concrete cover of 1.5 in., and use of a No. 3 stirrup. Situations requiring additional concrete cover, use of larger stirrups, or the intersection of beams with columns, may necessitate widths greater than those given in Table 7.1.

Beam Depth

For specification of the construction, the beam depth is defined by the overall concrete dimension: h in Fig. 7.6. For structural design, however, the critical depth dimension is that from the center of the tension reinforcing to the far side of the concrete: d in Fig. 7.6. While the selection of the depth is partly a matter of satisfying structural requirements, it is typically constrained by other considerations in the building design.

Figure 7.6 shows a section through a typical building floor/ceiling with a concrete slab-and-beam structure. In this situation

TABLE 7.1   Minimum Beam Widths[a]

Number of Bars	Bar Size								
	3	4	5	6	7	8	9	10	11
2	10	10	10	10	10	10	10	10	10
3	10	10	10	10	10	10	10	11	11
4	10	10	10	10	11	11	12	13	14
5	10	11	11	12	12	13	14	16	17
6	11	12	13	14	14	15	17	18	20

[a] Minimum width in inches for beams with 1.5-in. cover, No. 3 U-stirrups, clear spacing between bars of one bar diameter or minimum of 1.0 in. Minimum practical width for beam with No. 3 U-stirrups: 10 in.

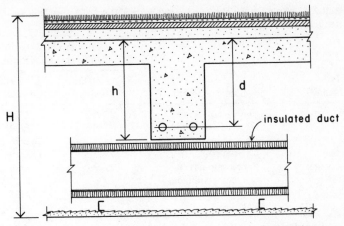

FIGURE 7.6 Concrete beam in typical multistory construction.

the critical depth from a general building design point of view is the overall thickness of the construction, shown as H in the illustration. In addition to the concrete structure, this includes allowances for the floor finish, the ceiling framing, and the passage of an insulated air duct. The net usable portion of H for the structure is shown as the dimension h, with the effective structural depth d being something less than h. Since the space defined by H is not highly usable for the building occupancy, there is a tendency to constrain it which works to limit the extravagant use of d.

Most concrete beams tend to fall within a limited range of terms of the ratio of width to depth. The typical range is for a width-to-depth ratio between $1:1.5$ and $1:2.5$, with an average of $1:2$. This is not a code requirement or a magic rule; it is merely the result of satisfying typical requirements for flexure, shear, bar spacing, and deflection.

Deflection Control

Deflection of spanning slabs and beams of poured-in-place concrete is controlled primarily by using recommended minimum thicknesses (overall height) expressed as a percentage of the span. Table 7.2 is adapted from a similar table given in Section 9.5

TABLE 7.2 Minimum Thickness of One-Way Slabs or Beams Unless Deflections Are Computed

Type of Member	End Conditions	Minimum Thickness of Slab or Height of Beam	
		$f_y = 40$ ksi [276 MPa]	$f_y = 60$ ksi [414 MPa]
Solid one-way slabs[a]	Simple support	$L/25$	$L/20$
	One end continuous	$L/30$	$L/24$
	Both ends continuous	$L/35$	$L/28$
	Cantilever	$L/12.5$	$L/10$
Beams or joists	Simple support	$L/20$	$L/16$
	One end continuous	$L/23$	$L/18.5$
	Both ends continuous	$L/26$	$L/21$
	Cantilever	$L/10$	$L/8$

Source: Data adapted from *Building Code Requirements for Reinforced Concrete* (ACI 318-89), 1989, ed., with permission of the publishers, American Concrete Institute.
[a] Valid only for members not supporting or attached to partitions or other construction likely to be damaged by large deflections.

of the ACI Code and yields minimum thicknesses as a fraction of the span. Table values apply only for concrete of normal weight (made with ordinary sand and gravel) and for reinforcing with f_y of 60 ksi [414 MPa]. The Code supplies correction factors for other concrete weights and reinforcing grades. The Code further stipulates that these recommendations apply only where beam deflections are not critical for other elements of the building construction, such as supported partitions subject to cracking caused by beam deflections.

Table 7.3 yields maximum spans for beams with various overall depths. These are based on the requirements given in Table 7.2. It should be noted that these are *limits* and are not necessarily practical or efficient values. Use of these limits will usually result in beams having a great amount of reinforcing, whereas economy is generally achieved by using minimum amounts of reinforcing.

Deflection of concrete structures presents a number of special problems. For concrete with ordinary reinforcing (not prestressed), flexural action normally results in tension cracking of

TABLE 7.3   Maximum Spans for Beams[a]

Overall Beam Depth, h (in.)	Maximum Permissible Span (ft)			
	Simply Supported	One End Continuous	Both Ends Continuous	Cantilever
10	13.3	15.4	17.5	6.7
12	16	18.5	21	8
14	18.7	21.6	24.5	9.3
16	21.3	24.7	28	10.7
18	24	27.7	31.5	12
20	26.7	30.8	35	13.3
24	32	37.0	42	16
30	40	46.2	52.5	20
36	48	55.5	63	24

[a] Based on requirements of Table 7.2. For normal-weight concrete and reinforcing with $f_y = 60$ ksi. For $f_y = 40$ ksi, multiply table values by 1.25.

the concrete at points of maximum bending. Thus the presence of cracks in the bottom of a beam at midspan points and in the top over supports is to be expected. In general, the size (and visibility) of these cracks will be proportionate to the amount of beam curvature produced by deflection. Crack size will also be greater for long spans and for deep beams. If visible cracking is considered objectionable, more conservative depth-to-span ratios should be used, especially for spans over 30 ft and beam depths over 30 in.

Creep of concrete (see Sec. 2.6) results in additional deflections over time. This is caused by the sustained loads—essentially the dead load of the construction. Deflection controls reflect concern for this as well as for the instantaneous deflection under live load, the latter being the major concern in structures of wood and steel.

In beams, deflections, especially creep deflections, may be reduced by the use of some compressive reinforcing. Where deflections are of concern, or where depth-to-span ratios are pushed to their limits, it is advisable to use some compressive reinforcing, consisting of continuous top bars.

When, for whatever reasons, deflections are deemed to be critical, computations of actual values of deflection may be necessary. Section 9.5 of the ACI Code provides directions for such computations; they are quite complex in most cases, and beyond the scope of this work. In actual design work, however, they are required very infrequently.

Design of Continuous Beams

Continuous beams are typically indeterminate and must be investigated for the bending moments and shears that are critical for the various loading conditions. When the beams are not involved in rigid-frame actions (as when they occur on column lines in multistory buildings), it may be possible to use approximate analysis methods, as described in Section 8.3 of the ACI Code (Ref. 1). An illustration of such a procedure is shown in the design of a concrete floor structure in Chapter 12.

In contrast to beams of wood and steel, those of concrete must be designed for the changing internal force conditions along their length. The single, maximum values for bending moment and shear may be critical in establishing the required beam size, but requirements for reinforcement must be investigated at all supports and midspan locations. A procedure for this is illustrated in the design examples in Chapter 12.

7.4 ONE-WAY JOIST CONSTRUCTION

Figure 7.7 shows a partial framing plan and some details for a type of construction that utilizes a series of very closely spaced beams and a relatively thin solid slab. Because of its resemblance to ordinary wood joist construction, this is called concrete joist construction. This system is generally the lightest (in dead weight) of any type of flat-spanning, poured-in-place concrete construction and is structurally well suited to the light loads and medium spans of office buildings and commercial retail buildings.

Slabs as thin as 2 in. and joists as narrow as 4 in. are used with this construction. Because of the thinness of the parts and the small amount of cover provided for reinforcement (typically $\frac{3}{4}$ to 1 in. for joists versus 1.5 in. for ordinary beams), the construction has very low resistance to fire, especially when exposed from the

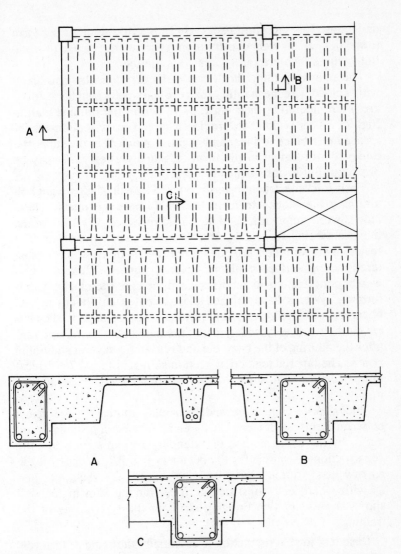

FIGURE 7.7 Framing layout plan for a typical one-way joist construction system.

underside. It is therefore necessary to provide some form of fire protection, as for steel construction, or to restrict its use to situations where high fire ratings are not required.

The relatively thin, short-span slabs are typically reinforced with welded wire mesh rather than ordinary deformed bars. Joists are often tapered at their ends, as shown in the framing plan in Fig. 7.7. This is done to provide a larger cross section for increased resistance to shear and negative moment at the supports. Shear reinforcement in the form of single vertical bars may be provided, but is not frequently used.

Early joist construction was produced by using lightweight hollow clay tile blocks to form the voids between joists. These blocks were simply arranged in spaced rows on top of the forms, the joists being formed by the spaces between the rows. The resulting construction provided a flat underside to which a plastered ceiling surface could be directly applied. Hollow, lightweight concrete blocks later replaced the clay tile blocks. Other forming systems have utilized plastic-coated cardboard boxes, fiberglass-reinforced pans, and formed sheet-metal pans. The latter method was very widely used, the metal pans being pried off after the pouring of the concrete and reused for several additional pours. The tapered joist cross section shown in Fig. 7.7 is typical of this construction, since the removal of the metal pans requires it.

Wider joists can be formed by simply increasing the space between forms, with large beams being formed in a similar manner or by the usual method of extending a beam stem below the construction, as shown for the beams in Fig. 7.7. Because of the narrow joist forms, cross-bridging is usually required, just as with wood joist construction. The framing plan in Fig. 7.7 shows the use of two bridging strips in the typical bay of the framing.

Design of joist construction is essentially the same as for ordinary slab-and-beam construction. Some special regulations are given in the ACI Code for this construction, such as the reduced cover mentioned previously. Because joists are so commonly formed with standard-sized metal forms, there are tabulated designs for typical systems in various handbooks. The *CRSI Handbook* (Ref. 3) has extensive tables offering complete designs for various spans, loadings, pan sizes, and so on. Whether for final design or simply for a quick preliminary design, the use of such tables is quite efficient.

One-way joist construction was highly popular in earlier times, but has become less utilized, due to its lack of fire resistance and the emergence of other systems. The popularity of lighter, less fire-resistive ceiling construction has been a contributing factor. In the right situation, however, it is still a highly efficient type of construction.

7.5 WAFFLE CONSTRUCTION

Waffle construction consists of two-way spanning joists that are formed in a manner similar to that for one-way spanning joists, using forming units of metal, plastic, or cardboard to produce the void spaces between the joists. The most widely used type of waffle construction is the waffle flat slab, in which solid portions around column supports are produced by omitting the void-making forms. An example of a portion of such a system is shown in Fig. 7.8. This type of system is analogous to the solid flat slab, which will be discussed in Sec. 7.6. At points of discontinuity in the plan—such as at large openings or at edges of the building—it is usually necessary to form beams. These beams may be produced as projections below the waffle, as shown in Fig. 7.8, or may be created within the waffle depth by omitting a row of the void-making forms, as shown in Fig. 7.9.

If beams are provided on all of the column lines, as shown in Fig. 7.9, the construction is analogous to the two-way solid slab with edge supports, as discussed in Sec. 7.6. With this system, the solid portions around the column are not required, since the waffle itself does not achieve the transfer of high shear or development of the high negative moments at the columns.

As with the one-way joist construction, fire ratings are low for ordinary waffle construction. The system is best suited for situations involving relatively light loads, medium-to-long spans, approximately square column bays, and a reasonable number of multiple bays in each direction.

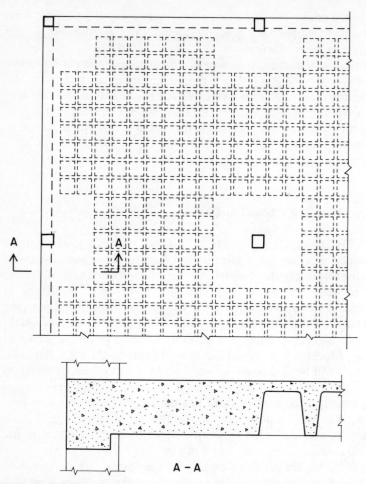

FIGURE 7.8 Framing layout plan for a waffle system without column-line beams.

For the waffle construction shown in Fig. 7.8, the edge of the structure represents a major discontinuity when the column supports occur immediately at the edge, as shown. Where planning permits, a more efficient use of the system is represented by the partial framing plan shown in Fig. 7.10, in which the edge occurs some distance past the columns. This projected edge provides a greater shear periphery around the column and helps to generate a negative moment, preserving the continuous character of the spanning structure. With the use of the projected edge, it may be possible to eliminate the edge beams shown in Fig. 7.8, thus preserving the waffle depth as a constant.

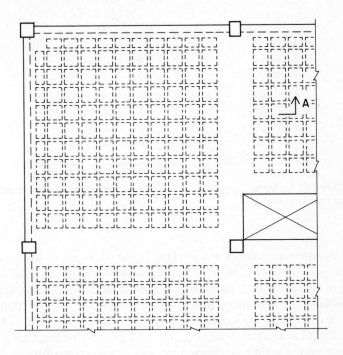

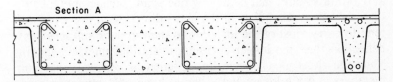

FIGURE 7.9 Framing layout plan for a waffle system with column-line beams.

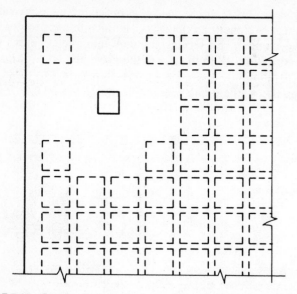

FIGURE 7.10 Framing layout plan for waffle system with cantilevered edges and no edge beams.

Another variation for the waffle is the blending of some one-way joist construction with the two-way waffle joists. This may be achieved by keeping the forming the same as for the rest of the waffle construction and merely using the ribs in one direction to create the spanning structure. One reason for doing this would be a situation similar to that shown in Fig. 7.9, where the large opening for a stair or elevator results in a portion of the waffle (the remainder of the bay containing the opening) being considerably out of square, that is, having one span considerably greater than the other. The joists in the short direction in this case will tend to carry most of the load due to their greater stiffness (less deflection than the longer spanning joists that intersect them). Thus the short joists would be designed as one-way spanning members and the longer joists would have only minimum reinforcing and serve as bridging elements.

The two-way spanning waffle systems are quite complex in structural behavior and their investigation and design are beyond the scope of this book. Some aspects of this work are discussed in the next section, since there are many similarities between the two-way spanning waffle systems and the two-way spanning solid slab systems. As with the one-way joist system, there are some tabulated designs in various handbooks that may be useful for either final or preliminary design. The *CRSI Handbook* (Ref. 3) mentioned previously has some such tables.

For all two-way construction, such as the waffle system, real feasible use of the system depends on some logic in terms of development of the general building plans regarding the arrangements of structural supports, locations of openings, length of spans, and so on. In the right situation, these systems may be able to realize their full potential, but if lack of order, symmetry, and other factors result in major adjustments away from the simple two-way functioning of the system, it may be very unreasonable to select such a structure. In some cases the waffle has been chosen strictly for its underside appearance, and has been pushed into use in situations not fitted to its nature. There may be some justifications for such a case, but the resulting structure is likely to be quite awkward.

7.6 TWO-WAY SPANNING SOLID-SLAB CONSTRUCTION

If reinforced in both directions, the solid concrete slab may span in two ways as well as one. The widest use of such a slab is in flat-slab or flat-plate construction. In flat-slab construction, beams are used only at points of discontinuity, with the typical system consisting only of the slab and the strengthening elements used at column supports. Typical details for a flat-slab system are shown in Fig. 7.11. Drop panels consisting of thickened portions square in plan are used to give additional resistance to the high shear and negative moment that develops at the column supports. Enlarged portions are also sometimes provided at the tops of the columns (called column capitals) to reduce the stresses in the slab further.

Two-way slab construction consists of multiple bays of solid two-way spanning slabs with edge supports consisting of bearing walls of concrete or masonry or of column-line beams formed in

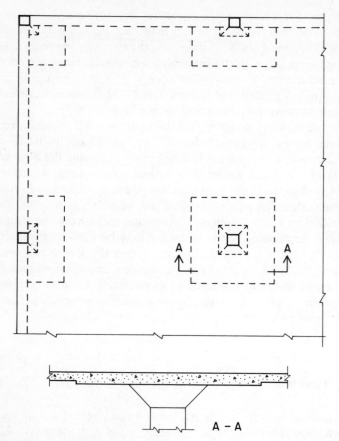

FIGURE 7.11 Framing layout plan for flat-slab construction with drop panels and column capitals.

the usual manner. Typical details for such a system are shown in Fig. 7.12.

Two-way solid-slab construction is generally favored over waffle construction where higher fire rating is required for the unprotected structure or where spans are short and loadings high. As with all types of two-way spanning systems, they function most efficiently where the spans in each direction are approximately the same.

For investigation and design, the flat slab (Fig. 7.11) is considered to consist of a series of one-way spanning solid-slab strips. Each of these strips spans through multiple bays in the manner of a continuous beam and is supported either by columns or by the strips that span in a direction perpendicular to it. The analogy for this is shown in Fig. 7.13a.

As shown in Fig. 7.13b, the slab strips are divided into two types: those passing over the columns, and those passing between columns, called middle strips. The complete structure con-

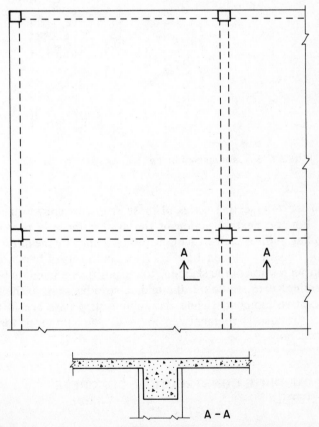

FIGURE 7.12 Framing layout plan for two-way spanning slab with edge supports.

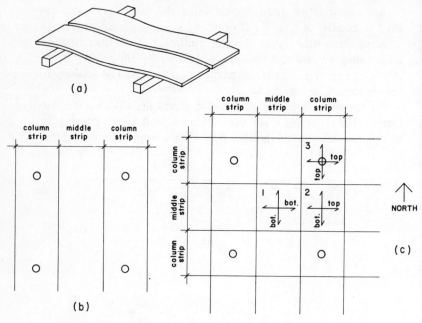

FIGURE 7.13 Development of the two-way spanning flat slab.

sists of the intersecting series of these strips, as shown in Fig. 7.13c. For the flexural action of the system there is two-way reinforcing in the slab at each of the boxes defined by the intersections of the strips. In box 1 in Fig. 7.13c, both sets of bars are in the bottom portion of the slab, due to the positive moment in both intersecting strips. In box 2, the middle-strip bars are in the top (for negative moment), while the column-strip bars are in the bottom (for positive moment). In box 3, the bars are in the top in both directions.

7.7 COMPOSITE CONSTRUCTION: CONCRETE PLUS STEEL

Figure 7.14 shows a section detail of a type of construction generally referred to as *composite construction*. This consists of a

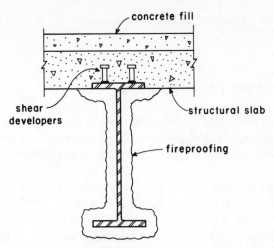

FIGURE 7.14 Composite construction: steel beams with cast-in-place concrete slab.

sitecast concrete spanning slab supported by structural steel beams, the two being made to interact by the use of shear developers welded to the top of the beams and embedded in the cast slab. The concrete may be formed by plywood sheets placed against the underside of the beam flange, resulting in the detail shown in Fig. 7.14.

A variation on the forming shown in Fig. 7.14 consists of using light-gage formed sheet steel decking to support the cast concrete. The shear developers are then site-welded through the thin deck to the tops of the beams.

Although it is common to refer to this form of construction as composite construction, in its true meaning the term covers any situation in which more than a single material is made to develop a singular structural response. By the general definition, therefore, even ordinary reinforced concrete is "composite," with the concrete interacting with the steel reinforcement. Other examples include laminated glazing (glass plus plastic), flitched beams (wood plus steel), and concrete fill on top of steel deck when the deck is bonded to the concrete.

The AISC Manual (*Manual of Steel Construction*, published by the American Institute of Steel Construction) contains data and design examples for the type of construction shown in Fig. 7.14.

7.8 PRECAST CONSTRUCTION

Precast construction components may be used to produce entire structural systems, but are more frequently used in combination with other structural components, such as the following:

1. *Precast Decks.* These may be solid slabs, hollow-cored slabs, or various contoured, ribbed slabs. They can be supported by sitecast concrete, masonry walls, or steel frames.
2. *Tilt-Up Walls.* These are routinely used with horizontal systems for roofs and floors consisting of wood or steel framing. However, they can also be combined with sitecast concrete or with other elements of precast concrete.

Flat-spanning systems can also be produced by using modular units of precast concrete to form sitecast systems, such as one-way joists or waffles. The exposed undersides of such construction is able to be achieved with finer detail and finish quality than with just about any other way of forming.

Some very imaginative structures have been developed with the use of custom-designed, complete systems of precast concrete. This generally demands a considerably greater design effort and considerable coordination of the design and production work. It also requires some real dedication of everyone involved to accomplish something way beyond the routine work of construction.

7.9 USE OF PRESTRESSING

There are various advantages to the use of prestressing in flat-spanning structures. A principal one is the reduction of creep deflections and a reduction of cracking; the latter being significant when the underside of the construction is exposed to view. Prestressing is used generally with flat-spanning precast concrete components, but can also be used with some sitecast systems in the form of post-tensioning.

General consideration for use of prestressing are discussed in Sec. 2.8.

7.10   USE OF DESIGN AIDS

The design of various elements of reinforced concrete can be aided—or in many cases totally achieved—by the use of various prepared materials. Handbooks (see Ref. 3) present complete data for various elements, such as footings, columns, one-way slabs, joist construction, waffle systems, and two-way slab systems. For the design of a single footing or a one-way slab, the handbook merely represents a convenience, or a shortcut, to a final design. For columns subjected to bending, for waffle construction, and for two-way slab systems, "longhand" design (without aid other than that from a pocket calculator) is really not feasible. In the latter cases, handbook data may be used to establish a reasonable preliminary design, which may then be customfit to the specific conditions by some investigation and computations. Even the largest of handbooks cannot present all possible combinations of values of f'_c, grade of reinforcing bars, value of superimposed loads, and so on. Thus only coincidentally will handbook data be exactly correct for any specific design job.

In the age of the computer, there is a considerable array of software available for the routine tasks of structural design. For many of the complex and laborious problems of design of reinforced concrete structures, these are a real boon for anyone able to utilize them.

8

CONCRETE COLUMNS

Concrete columns occur most often as reinforced members, with sitecast concrete and steel reinforcement sharing compressive loads and working as in beams to develop any required bending resistance. Common forms consist of round, square, or oblong rectangular cross sections of solid concrete with steel bars placed as close as possible to the column perimeter. With the normal height of columns, the steel bars on their own are quite slender, and to prevent their buckling out through the thin concrete cover, some form of restraint is used to tie them back to the column core.

Computations for column design are now quite complex, and are commonly done with computer-aided processes in professional practice. Preliminary design work may utilize some design aids in the form of handbook tabulations, such as those given in the *CRSI Handbook* (Ref. 3) or materials available from the American Concrete Institute (ACI) or Portland Cement Association (PCA). However, with the range of possible variables—including concrete strength, steel yield strength, column type, column shape, and column dimensions—it is difficult to provide adequate tabulations that do not become prohibitively voluminous.

Materials in this chapter are limited to a general discussion of the functions of columns, the general problems of design, detailing, and construction, and the general cases for design. To permit some simple demonstrations of design examples, some data are provided for a working stress method design of some ordinary columns.

8.1 TYPES OF COLUMNS

Concrete columns occur most often as the vertical support elements in a structure generally built of poured-in-place concrete. This is the situation discussed in this chapter. Columns may also occur as precast elements; the general concerns of the precast structure are discussed in Sec. 3.10. Very short columns, called *pedestals*, are sometimes used in the support system for columns or other structures. The ordinary pedestal is discussed as a foundation transitional device in Chapter 9. Walls that serve as vertical compression supports (called *bearing walls)* are discussed in Chapter 10.

The cast-in-place concrete column usually falls into one of the following categories:

1. Square columns with tied reinforcing.
2. Oblong columns with tied reinforcing.
3. Round columns with tied reinforcing.
4. Round columns with spiral-bound reinforcing.
5. Square columns with spiral-bound reinforcing.
6. Columns of other geometries (L- or T-shaped, octagonal, etc.) with either tied or spiral-bound reinforcing.

Obviously, the choice of column cross-sectional shape is an architectural as well as a structural decision. However, forming methods and costs, arrangement and installation of reinforcing,

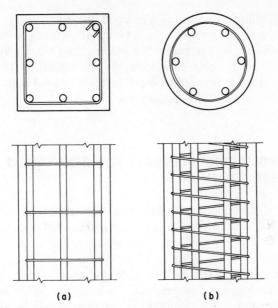

FIGURE 8.1 Typical reinforced concrete columns: (*a*) with loop ties; (*b*) with a spiral wrap.

and relations of the column form and dimensions to other parts of the structural system must also be dealt with.

In tied columns the longitudinal reinforcing is held in place by loop ties made of small-diameter reinforcing bars, commonly No. 3 or No. 4. Such a column is represented by the square section shown in Fig. 8.1*a*. This type of reinforcing can quite readily accommodate other geometries as well as the square. The design of such a column is discussed in Sec. 8.8.

Spiral columns are those in which the longitudinal reinforcing is placed in a circle, with the whole group of bars enclosed by a continuous cylindrical spiral made from steel rod or large-diameter steel wire. Although this reinforcing system obviously works best with a round column section, it can be used also with other geometries. A round column of this type is shown in Fig. 8.1*b*.

Experience has shown the spiral column to be slightly stronger than an equivalent tied column with the same amount of concrete and reinforcing. For this reason, code provisions allow slightly more load on spiral columns. Spiral reinforcing tends to be expensive, however, and the round bar pattern does not always mesh well with other construction details in buildings. Thus tied columns are often favored where restrictions on the outer dimensions of the sections are not severe.

8.2 GENERAL REQUIREMENTS FOR REINFORCED CONCRETE COLUMNS

Code provisions and practical construction considerations place a number of restrictions on column dimensions and choice of reinforcing.

Column Size. The current code does not contain limits for column dimensions. For practical reasons, the following limits are recommended. Rectangular tied columns should be limited to a minimum area of 100 in.^{2} and a side dimension of 10 in. if square and 8 in. if oblong. Spiral columns should be limited to a minimum size of 12 in. if either round or square.

Reinforcing. Minimum bar size is No. 5. The minimum number of bars is four for tied columns, five for spiral columns. The minimum amount of area of steel is 1% of the gross column area. A maximum area of steel of 8% of the gross area is permitted, but bar spacing limitations makes this difficult to achieve; 4% is a more practical limit. The 1989 ACI Code stipulates that for a compression member with a larger cross section than required by considerations of loading, a reduced effective area not less than one-half the total area may be used to determine minimum reinforcement and design strength.

Ties. Ties shall be at least No. 3 for bars No. 10 and smaller. No. 4 ties should be used for bars that are No. 11 and larger. Vertical spacing of ties shall be not more than 16 times the bar diameter, 48 times the tie diameter, or the least dimension of the column. Ties shall be arranged so that every corner and alternate longitudinal bar is held by the corner of a tie with an included angle of not greater than 135°, and no bar shall be farther than 6 in. clear from such a supported bar. Complete circular ties may be used for bars placed in a circular pattern.

Concrete Cover. A minimum of 1.5 in. is needed when the column surface is not exposed to weather or in contact with the ground; 2 in. should be used for formed surfaces exposed to the weather or in contact with ground; 3 in. are necessary if the concrete is cast against earth.

Spacing of Bars. Clear distance between bars shall not be less than 1.5 times the bar diameter, 1.33 times the maximum specified size for the coarse aggregate, or 1.5 in.

8.3 COLUMNS WITH AXIAL COMPRESSION PLUS BENDING

Due to the nature of most concrete structures, current design practices generally do not consider the possibility of a concrete column with axial compression alone. That is to say, the existence of some bending moment is always considered together with the axial force. Figure 8.2 illustrates the nature of the so-called *interaction response* for a concrete column, with a range of combinations of axial load plus bending moment. In general, there are three basic ranges of this behavior, as follows (see the dashed lines in Fig. 8.2):

1. *Large Axial Force, Minor Moment.* For this case the moment has little effect, and the resistance to pure axial force is only negligibly reduced.
2. *Significant Values for Both Axial Force and Moment.* For this case the analysis for design must include the full combined force effects, that is, the interaction of the axial force and the bending moment.
3. *Large Bending Moment, Minor Axial Force.* For this case the column behaves essentially as a doubly reinforced (ten-

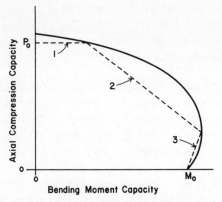

FIGURE 8.2 Interaction of axial compression and bending in a reinforced concrete column: solid line indicates typical response; dashed line indicates separate zones of response forms.

sion and compression reinforced) member, with its capacity for moment resistance affected only slightly by the axial force.

In Fig. 8.2 the solid line on the graph represents the true response of the column—a form of behavior verified by many load tests on laboratory specimens. The dashed line on the graph represents the generalization of the three types of response just described.

The terminal points of the interaction response—pure axial compression or pure bending moment—may be reasonably easily determined (P_o and M_o in Fig. 8.2). The interaction responses between these two limits require complex analyses beyond the scope of this book.

8.4 COLUMNS IN FRAMES

Reinforced concrete columns seldom occur as single, pin-ended members, as opposed to most wood columns and many steel columns. This condition may exist for some precast concrete

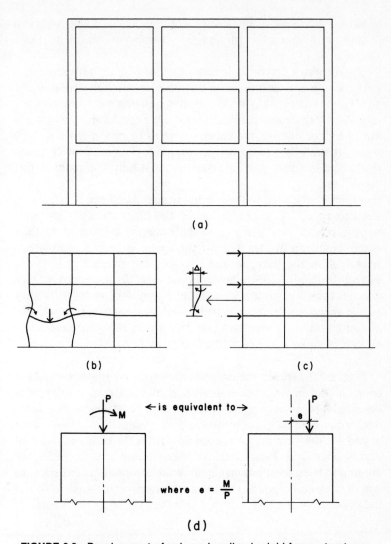

FIGURE 8.3 Development of column bending in rigid frame structures.

columns, but almost all sitecast columns occur as members in frames, with interaction of the frame members in the manner of a so-called *rigid frame*.

Rigid frames derive their name from the joints between members, which are assumed to be *moment-resistive* (rotationally rigid) and thus capable of transmitting bending moments between the ends of the connected members. This condition may be visualized by considering the entire frame as being cut from a single piece of material, as shown in Fig. 8.3. The sitecast concrete frame and the all-welded steel frame most fully approximate this condition.

When the horizontal-spanning frame members (beams) are subjected to vertical gravity loads, the inclination of their ends to rotate transmits bending to the columns connected to their ends, as shown in Fig. 8.3*b*. If the frame is subjected to lateral loads (often the case, as rigid frames are frequently used for lateral bracing), the relative horizontal displacement of the column tops and bottoms (called lateral drift) transmits bending to the members connected to the columns (see Fig. 8.3*c*). The combinations of these loadings results in the general case of combined axial loads plus bending in all the members of a rigid frame.

Figure 8.3*d* shows the case for the effect on the cross section of a column in a frame: a condition of axial compression plus bending. For some purposes, it is useful to visualize this as an analogous eccentric compressive force, with the bending produced by the product of the compression times its distance of eccentricity (*e* in Fig. 8.3*d*). It is thus possible to consider the column to have a maximum capacity for compression (with $e = 0$) which is steadily reduced as the eccentricity is increased. This is the concept of the interaction graph (see Fig. 8.2).

8.5   MULTISTORY COLUMNS

Concrete columns occur frequently in multistory structures. In the sitecast structure, separate stories are typically cast in separate casting, with a cold joint (construction joint) between the successive pours. While this makes for a form of discontinuity, it does not significantly reduce the effective monolithic nature of the framed structure. Compression is continuous by the simple stacking of the levels of the heavy structure, and splicing of the reinforcement develops a form of tension continuity, permitting development of bending moments.

The typical arrangement of reinforcement in multistory columns is shown in Fig. 6.29, which illustrates the form of bar development required to achieve the splicing of the reinforcement. This is essentially compressive reinforcement, so its development is viewed in those terms. However, an important practical function of the column bars is simply to tie the structure together through the discontinuous construction joints.

Load conditions change in successive stories of the multistory structure. It is therefore common to change both the column size and the reinforcement. Design considerations for this are discussed in the examples in Chapter 12.

In very tall structures the magnitude of compression in lower stories requires columns with very high resistance. There is often some practical limit to column sizes, so that all efforts are made to obtain strength increases other than by simply increasing the mass of concrete. The three basic means of achieving this are:

1. Increase the amount of reinforcement, packing columns with the maximum amount that is feasible and allowable by codes.
2. Increase the yield strength of the steel, using as much as twice the strength for ordinary bars.
3. Increase the strength of the concrete.

The super strength column is a clear case for use of the highest achievable concrete strengths, and is indeed the application that has resulted recently in spiraling high values for design strength. Strengths as high as 20,000 psi have already been achieved, and higher ones are being proposed.

8.6 CONSIDERATIONS FOR COLUMN FORM

Usually, a number of possible combinations of reinforcing bars may be assembled to satisfy the steel area requirement for a given column. Aside from providing for the area, the number of bars must also work reasonably in the layout of the column. Figure 8.4 shows a number of tied columns with various number of bars. When a column is small, the preferred choice is usually that of the simple four-bar layout, with one bar in each corner and a single peripheral tie. As the column gets larger, the distance between the corner bars gets larger, and it is best to use more bars so that the reinforcing is spread out around the column periphery. For a symmetrical layout and the simplest of tie layouts, the best choice is for numbers that are multiples of four, as shown in Fig. 8.4a. The number of additional ties required for these layouts depends on the size of the column and the considerations discussed in Sec. 8.2.

An unsymmetrical bar arrangement is not necessarily bad, even though the column and its construction details are otherwise not oriented differently on the two axes. In situations where moments may be greater on one axis, the unsymmetrical layout is actually preferred; in fact, the column shape will also be more effective if it is unsymmetrical, as shown for the oblong shapes in Fig. 8.4c.

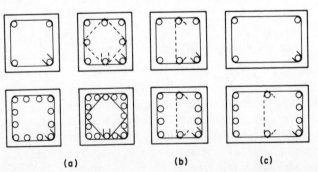

(a)   (b)   (c)

FIGURE 8.4 Typical bar placement and tie patterns for tied columns.

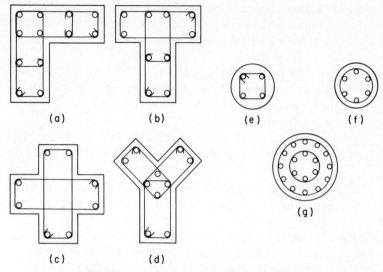

FIGURE 8.5 Tied columns of various shapes.

Figure 8.5 shows a number of special column shapes developed as tied columns. Although spirals could be used in some cases for such shapes, the use of ties allows much greater flexibility and simplicity of construction. One reason for using ties may be the column dimensions; there being a practical lower limit of about 12 in. in width for a spiral-bound column.

Round columns are most often formed as shown in Fig. 8.5e, if built as tied columns. This allows for a minimum reinforcing with four bars. If a round pattern is used (as it must be for a spiral-bound column), the usual minimum number recommended is six bars. Spacing of bars is much more critical in spiral-bound circular arrangements, making if very difficult to use high percentages of steel in the column section. For very large diameter columns it is possible to use sets of concentric spirals, as shown in Fig. 8.5g.

For cast-in-place columns a concern that must be dealt with is that for vertical splicing of the steel bars. As shown in Fig. 6.29, the two places where this commonly occurs are at the top of the foundation and at floors where a multistory column continues upward. At these points there are three ways to achieve the vertical continuity (splicing) of the steel bars, any of which may be appropriate for a given situation.

1. Bars may be lapped the required distance for development of the compression splice, as shown in Fig. 6.29. For bars of smaller dimension and lower yield strengths, this is usually the desired method.
2. Bars may have milled square-cut ends butted together with a grasping device to prevent separation in a horizontal direction.
3. Bars may be welded with full-penetration butt welds or by welding of the grasping device described for method 2.

The choice of splicing methods is basically a matter of cost comparison, but is also affected by the size of the bars, the degree of concern for bar spacing in the column arrangement, and possibly for a need for some development of tension through the splice if uplift or high magnitudes of moments exist. If lapped splicing is used, a problem that must be considered is the bar layout at the location of the splice, at which point there will be twice the usual number of bars. The lapped bars may be adjacent to each other, but the usual considerations for space between bars must be made. If spacing is not critical, the arrangement shown in Fig. 8.6a is usually chosen, with the spliced sets of bars next to each other at the tie perimeter. If spacing does not permit the arrangement in Fig. 8.6a, that shown in Fig. 8.6b may be used, with the

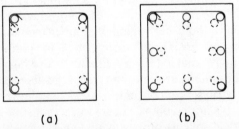

(a)   (b)

FIGURE 8.6 Bar placement at splice locations.

lapped sets in concentric patterns. The latter arrangement is commonly used for spiral-bound columns, where spacing is often critical.

8.7 DESIGN METHODS

At the present, design of concrete columns is mostly achieved by using either tabulations from handbooks or computer-aided procedures. The present ACI Code (ACI 318-89) does not permit design of columns by working stress methods in a direct manner; stipulating instead that a capacity of 40% of that determined by strength methods be used if working stress procedures are used in design. Using the code formulas and requirements to design by "hand operation" with both axial compression and bending present at all times is prohibitively laborious. The number of variables present (column shape and size, f'_c, f_y, number and size of bars, arrangement of bars, etc.) adds to the usual problems of column design to make for a situation much worse than those for wood or steel columns.

The large number of variables also works against the efficiency of handbook tables. Even if a single concrete strength (f'_c) and a single steel yield strength (f_y) are used, tables would be very extensive if all sizes, shapes, and types (tied and spiral) of columns were included. Even with a very limited range of variables, handbook tables are much larger than those for wood or steel columns. They are, nevertheless, often quite useful for preliminary design estimation of column sizes.

The obvious preference when relationships are complex, requirements are tedious and extensive, and there are a large number of variables, is for a computer-aided system. It is hard to imagine a professional design office that is turning out designs of concrete structures on a regular basis at the present without computer-aided methods. We will not attempt to explain these methods, but the reader should be aware that the software required for them is readily available.

As in other situations, the common practices at any given time tend to narrow down to a limited usage of any type of construction, even though the potential for variation is extensive. It is thus possible to use some very limited but easy-to-use design aids to make early selections for design. These approximations may be adequate for preliminary building planning, cost estimates, and some preliminary structural analyses.

One highly useful reference is the *CRSI Handbook* (Ref. 3), which contains quite extensive tables for design of both tied and spiral columns. Square, round, and some oblong shapes, plus some range of concrete and steel strengths are included. Table format uses the equivalent eccentric load technique which is discussed in the next section.

8.8 APPROXIMATE DESIGN OF TIED COLUMNS

Tied columns are much preferred due to the relative simplicity and usually lower cost of their construction, plus their adaptability to various column shapes (T, L, etc.) Even round columns— most naturally formed with spiral-bound reinforcing—are often made with ties instead, when the structural demands are modest. An exception to this is the situation of columns in rigid-frame structures in zones of high seismic risk, where the toughness of spiral columns is much preferred.

The column with moment is often designed using the equivalent eccentric load method. The general case for this is discussed in Sec. 8.3. The method consists of translating a compression plus bending situation into an equivalent one with an eccentric load, the moment becoming the product of the load and the eccentricity, as shown in Fig. 8.3d. This method is often used in presentation of tabular data for column capacities. It is also used in the development of the graphs in Figs. 8.7 and 8.8, which yield safe service load capacities for simple square and round tied columns of concrete with $f'_c = 4$ ksi [27.6 MPa] and reinforcing with $f_y = 60$ ksi [414 MPa].

Figure 8.7 gives safe loads for a selected number of sizes of square tied columns. Loads are given for various degrees of eccentricity, which is a means for expressing axial load and bending

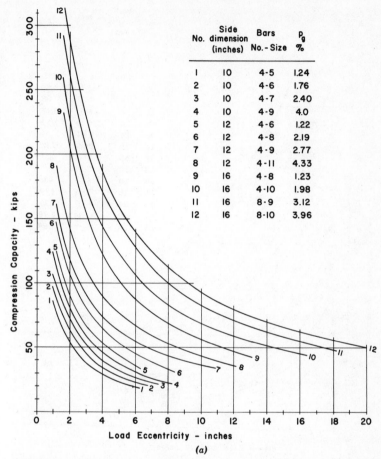

No.	Side dimension (inches)	Bars No. - Size	p_g %
1	10	4-5	1.24
2	10	4-6	1.76
3	10	4-7	2.40
4	10	4-9	4.0
5	12	4-6	1.22
6	12	4-8	2.19
7	12	4-9	2.77
8	12	4-11	4.33
9	16	4-8	1.23
10	16	4-10	1.98
11	16	8-9	3.12
12	16	8-10	3.96

(a)

FIGURE 8.7 Safe service loads for square tied columns with $f'_c = 4$ ksi and $f_y = 60$ ksi.

moment combinations. The computed moment on the column is translated into an equivalent eccentric loading, as shown in Fig. 8.3*d*. Data for the curves were computed by using 40% of the load determined by strength design methods, as required by the 1989 ACI Code.

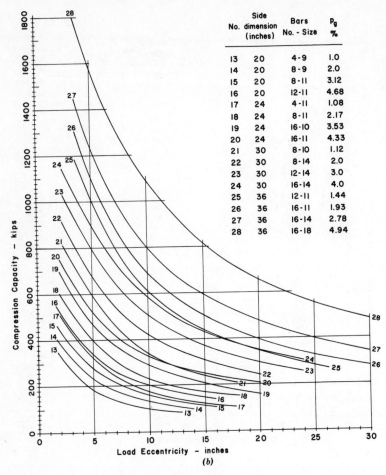

No.	Side dimension (inches)	Bars No. - Size	P_g %
13	20	4 - 9	1.0
14	20	8 - 9	2.0
15	20	8 - 11	3.12
16	20	12 - 11	4.68
17	24	4 - 11	1.08
18	24	8 - 11	2.17
19	24	16 - 10	3.53
20	24	16 - 11	4.33
21	30	8 - 10	1.12
22	30	8 - 14	2.0
23	30	12 - 14	3.0
24	30	16 - 14	4.0
25	36	12 - 11	1.44
26	36	16 - 11	1.93
27	36	16 - 14	2.78
28	36	16 - 18	4.94

FIGURE 8.7 (Continued)

The following examples illustrate the use of Fig. 8.7 for the design of tied columns.

Example 1. A column with $f'_c = 4$ ksi and steel with $f_y = 60$ ksi sustains an axial compression load of 400 kips. Find the minimum

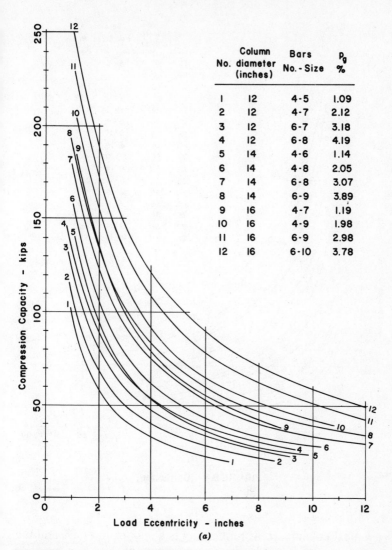

No.	Column diameter (inches)	Bars No.-Size	p_g %
1	12	4-5	1.09
2	12	4-7	2.12
3	12	6-7	3.18
4	12	6-8	4.19
5	14	4-6	1.14
6	14	4-8	2.05
7	14	6-8	3.07
8	14	6-9	3.89
9	16	4-7	1.19
10	16	4-9	1.98
11	16	6-9	2.98
12	16	6-10	3.78

FIGURE 8.8 Safe service loads for round tied columns with $f'_c = 4$ ksi and $f_y = 60$ ksi.

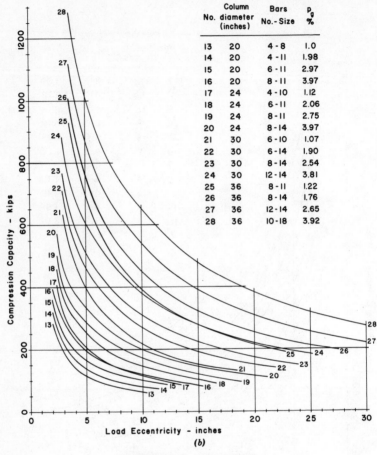

No.	Column diameter (inches)	Bars No.-Size	p_g %
13	20	4 - 8	1.0
14	20	4 - 11	1.98
15	20	6 - 11	2.97
16	20	8 - 11	3.97
17	24	4 - 10	1.12
18	24	6 - 11	2.06
19	24	8 - 11	2.75
20	24	8 - 14	3.97
21	30	6 - 10	1.07
22	30	6 - 14	1.90
23	30	8 - 14	2.54
24	30	12 - 14	3.81
25	36	8 - 11	1.22
26	36	8 - 14	1.76
27	36	12 - 14	2.65
28	36	10 - 18	3.92

FIGURE 8.8 (Continued)

practical column size if reinforcing is a maximum of 4% and the maximum size if reinforcing is a minimum of 1%.

Solution: Using Fig. 8.7a, we find from the sizes given:

Minimum column is 20 in. square with 8 No. 9 (curve 14).
Maximum capacity is 410 kips, $p_g = 2.0\%$.

Maximum size is 24 in. square with 4 No. 11 (curve 17).
Maximum capacity is 510 kips, $p_g = 1.08\%$.

It should be apparent that it is possible to use an 18- or 19-in.
column as the minimum size and to use a 22- or 23-in. column as
the maximum size. Since these sizes are not given in the figure,
we cannot verify them for certain without using strength design
procedures.

Example 2. A square tied column with $f'_c = 4$ ksi and steel with
$f_y = 60$ ksi sustains an axial load of 400 kips and a bending moment of 200 kip-ft. Determine the minimum size column and its
reinforcing.

Solution: We first determine the equivalent eccentricity, as
shown in Fig. 8.3*d*. Thus

$$e = \frac{M}{P} = \frac{200 \times 12}{400} = 6 \text{ in.}$$

Then, from Fig. 8.7*b*, we find:

Minimum size is 24 in. square with 16 No. 10 bars.
Capacity at 6-in. eccentricity is 410 kips.

PROBLEM 8.8.A,B,C,D,E. Using Fig. 8.7, pick the minimum size
square tied column and its reinforcing for the following combinations of axial load and bending moment:

	Axial Compressive Load (kips)	Bending Moment (kip-ft)
A	100	25
B	100	50
C	150	75
D	200	100
E	300	150

8.9 ROUND COLUMNS

Round columns may be designed and built as spiral columns, as described in Sec. 8.1, or they may be developed as tied columns with the bars placed in a circle and held by a series of round circumferential ties. Because of the cost of spirals, it is usually more economical to use the tied column, so it is often used unless the additional strength or other behavioral characteristics of the spiral column are required.

It is also possible to use rectangular bar layouts and tie patterns as shown in Fig. 8.5e inside a round column form. In such cases, the column is usually designed as a square column using the square shape that can be included within the round form. It is thus possible to use a four-bar column for small-diameter, round column forms.

Figure 8.8 gives safe loads for round columns that are designed as tied columns. Load values have been adapted from values determined by strength design methods. The curves in Fig. 8.8 are similar to those for the square columns in Fig. 8.7, and their use is similar to that demonstrated in Examples 2 and 3.

PROBLEM 8.9.A,B,C,D,E. Using Fig. 8.8, pick the minimum size round column and its reinforcing for the load and moment combinations in Problem 8.8.

8.10 COLUMN SLENDERNESS

Cast-in-place concrete columns tend to be quite stout in profile, so that slenderness is much less often a critical concern than with columns of wood or steel. The Code provides for consideration of slenderness, but permits the issue to be neglected when the L/r of the column falls below a controlled value. For rectangular columns this usually means that the effect is neglected when the ratio of unsupported height to side dimension is less than about 12. This is roughly analogous to the case for the wood column with L/D less than 11.

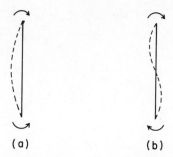

FIGURE 8.9 Conditions of bending for consideration of slenderness.

Slenderness effects must also be related to the conditions of bending for the column. Since bending is usually induced at the column ends, the two typical cases are those shown in Fig. 8.9. If two equal end moments, as shown in Fig. 8.9a, exist, the buckling effect is magnified, the P-delta effect is maximum, and the Code limits slenderness without reduction to L/d ratios of 6.6 or less. The condition in Fig. 8.9a is not the common case, however, the more typical condition in framed structures being that shown in Fig. 8.9b, for which the L/d limit for equal end moments jumps to 13.8 before reduction for slenderness is required.

When slenderness must be considered, the complex procedures required are simply built into your friendly neighborhood software program. One should be aware, however, that reduction for slenderness is not considered in the usual design aids, such as tables or graphs.

9

FOUNDATIONS

Almost every building utilizes some concrete construction that is built directly in contact with the ground. Such elements may include the following:

Shallow bearing footings, consisting of concrete pads that are used to spread the vertical loads of the building onto the supporting soil materials.

Concrete piles, pile caps, or concrete-filled excavated shafts used to develop deep foundations.

Concrete walls, enclosing below-grade spaces or forming grade beams.

Concrete pavements, for driveways, walks, parking lots, patios, or building floors placed directly on the ground.

Retaining walls, used to achieve sudden changes of the soil profile.

Underground tunnels and vaults for service systems.

Bases for elevators and other equipment.

This construction constitutes a major usage of concrete. Its low bulk cost, general nonrotting nature, rock-like character, general overall stiffness of thick elements, and ease of placement in the ground-level work make it the natural choice for most work. In days past, much of this construction was achieved with masonry materials. This is still possible, but now generally limited to walls in special situations.

This is potentially a large topic, if fully developed. Actually, there are two topics involved: concrete design and soil mechanics. The presentations in this chapter are limited to simple bearing-type foundation elements. Soil properties and usage are discussed briefly, as related to the bearing foundation situations. For a more extensive discussion of the topic, the reader is referred to *Simplified Design of Building Foundations* (Ref. 13).

9.1 GENERAL CONCERNS FOR FOUNDATIONS

The design of the foundation for a building cannot be separated from the overall problems of the building structure and the building and site designs in general. Nevertheless, it is useful to consider the specific aspects of the foundation design that must be dealt with.

Site Exploration

For purposes of the foundation design, as well as for the building and site development in general, it is necessary to know the actual site conditions. This investigation usually consists of two parts: determination of the ground surface conditions, and of the subsurface conditions. The surface conditions are determined by a site survey that establishes the three-dimensional geometry of the surface and the location of various objects and features on the site. Where they exist, the location of buried objects such as sewer lines, underground power and telephone lines, and so on, may also be shown on the site survey.

Unless they are known from previous explorations, the subsurface conditions must be determined by penetrating the surface to obtain samples of materials at various levels below the surface.

Inspection and testing of these samples in the field, and possibly in a testing lab, is used to identify the materials and to establish a general description of the subsurface conditions.

Site Design

Site design consists of positioning the building on the site and the general development, or redevelopment, of the site contours and features. The building must be both horizontally and vertically located. Recontouring the site may involve both taking away existing material (called *cutting*) and building up to a new surface with materials brought in or borrowed from other locations on the site (called *filling*). Development of controlled site drainage for water run-off is an important part of the site design.

Selection of Foundation Type

The first formal part of the foundation design is the determination of the type of foundation system to be used. This decision cannot normally be made until the surface and subsurface conditions are known in some detail and the general size, shape, and location of the building are determined. In some cases it may be necessary to proceed with an approximate design of several possible foundation schemes so that the results can be compared.

Design of Foundation Elements

With the building and site designs reasonably established, the site conditions known, and the type of foundation determined, work can proceed to the detailed design of individual structural elements of the foundation system.

9.2 SOIL CONDITIONS RELATED TO FOUNDATION DESIGN

The principal properties and behavior characteristics of soils that are of direct concern in foundation design are the following:

Strength. For bearing-type foundations the main concern is resistance to vertical compression. Resistance to horizontal pressure and to friction are of concern when foundations must resist the lateral forces of wind, earthquakes, or retained earth.

Strain Resistance. Deformation of soil under stress is of concern in designing for limitations of the movements of foundations, such as the vertical settlement of bearing foundations.

Stability. Frost action, fluctuations in water content, seismic shock, organic decomposition, and disturbance during construction are some of the things that may produce changes in physical properties of soils. The degree of sensitivity of the soil to these actions is called its relative stability.

Properties Affecting Construction Activity

A number of possible factors may affect construction activity, including the following:

The relative ease of excavation.

Ease of and possible effects of site dewatering during construction.

Feasibility of using excavated materials as fill material.

Ability of the soil to stand on a vertical side of an excavation.

Effects of construction activity—notably the movement of workers and equipment—on unstable soils.

Miscellaneous Conditions

In specific situations various factors may affect the foundation design or the problems to be dealt with during construction. Some examples are the following:

Location of the water table, affecting soil strength or stability, need for waterproofing basements, requirement for dewatering during construction, and so on.

Nonuniform soil conditions on the site, such as soil strata that are not horizontal, strips or pockets of poor soil, and so on.

Local frost conditions, affecting the depth required for bearing foundations and possible heave and settlement of exterior pavements.

Deep excavation or dewatering operations, possibly affecting the stability of adjacent properties, buildings, streets, and so on.

All of these concerns must be anticipated and dealt with in designing buildings and in planning and estimating construction costs. Persons charged with responsibility for design and planning foundation construction must have some understanding of the characteristics of ordinary soils so that they can translate information about site conditions into usable data.

9.3 FOUNDATION DESIGN CRITERIA

For the design of ordinary bearing-type foundations several structural properties of a soil must be established. The principal values are the following:

Allowable Bearing Pressure.   This is the maximum permissible value for vertical compression stress at the contact surface of bearing elements. It is typically quoted in units of pounds or kips per square foot of contact surface.

Compressibility.   This is the predicted amount of volumetric consolidation that determines the amount of settlement of the foundation. Quantification is usually done in terms of the actual dimension of vertical settlement predicted for the foundation.

Active Lateral Pressure.   This is the horizontal pressure exerted against retaining structures, visualized in its simplest form as an equivalent fluid pressure. Quantification is in terms of a density for the equivalent fluid given in actual unit weight value or as a percentage of the soil unit weight.

TABLE 9.1 Allowable Foundation and Lateral Pressure

CLASS OF MATERIALS[2]	ALLOWABLE FOUNDATION PRESSURE LBS. SQ. FT.[3]	LATERAL BEARING LBS./SQ. FT./ FT. OF DEPTH BELOW NATURAL GRADE[4]	LATERAL SLIDING[1]	
			COEF- FICIENT[5]	RESISTANCE LBS./SQ. FT.[6]
1. Massive Crystalline Bedrock	4000	1200	.79	
2. Sedimentary and Foliated Rock	2000	400	.35	
3. Sandy Gravel and/or Gravel (GW and GP)	2000	200	.35	
4. Sand, Silty Sand, Clayey Sand, Silty Gravel and Clayey Gravel (SW, SP, SM, SC, GM and GC)	1500	150	.25	
5. Clay, Sandy Clay, Silty Clay and Clayey Silt (CL, ML, MH and CH)	1000[7]	100		130

[1]Lateral bearing and lateral sliding resistance may be combined.

[2]For soil classifications OL, OH and PT (i.e., organic clays and peat), a foundation investigation shall be required.

[3]All values of allowable foundation pressure are for footings having a minimum width of 12 inches and a minimum depth of 12 inches into natural grade. Except as in Footnote 7 below, increase of 20 percent allowed for each additional foot of width and/or depth to a maximum value of three times the designated value.

[4]May be increased the amount of the designated value for each additional foot of depth to a maximum of 15 times the designated value. Isolated poles for uses such as flagpoles or signs and poles used to support buildings which are not adversely affected by a ½-inch motion at ground surface due to short-term lateral loads may be designed using lateral bearing values equal to two times the tabulated values.

[5]Coefficient to be multiplied by the dead load.

[6]Lateral sliding resistance value to be multiplied by the contact area. In no case shall the lateral sliding resistance exceed one half the dead load.

[7]No increase for width is allowed.

Source: Table No. 29-B from the *Uniform Building Code* (Ref. 4); reproduced with permission of the publishers, International Conference of Building Officials.

Passive Lateral Pressure.   This is the horizontal resistance offered by the soil to forces against the soil mass. It is also visualized as varying linearly with depth in the manner of a fluid pressure. Quantification is usually in terms of a specific pressure increase per unit of depth.

TABLE 9.2 Foundations for Stud Bearing Walls[1,2]

NUMBER OF FLOORS SUPPORTED BY THE FOUNDATION[3]	THICKNESS OF FOUNDATION WALL (Inches)		WIDTH OF FOOTING (Inches)	THICKNESS OF FOOTING (Inches)	DEPTH BELOW UNDISTURBED GROUND SURFACE (Inches)
	CONCRETE	UNIT MASONRY			
1	6	6	12	6	12
2	8	8	15	7	18
3	10	10	18	8	24

[1]Where unusual conditions or frost conditions are found, footings and foundations shall be as required in Section 2907 (a).

[2]The ground under the floor may be excavated to the elevation of the top of the footing.

[3]Foundations may support a roof in addition to the stipulated number of floors. Foundations supporting roofs only shall be as required for supporting one floor.

Source: Table No. 29-A from the *Uniform Building Code* (Ref. 4), reproduced with permission of the publishers, International Conference of Building Officials.

Friction Resistance.   This is the resistance to sliding along the contact bearing face of a footing. For cohesionless soils it is usually given as a friction coefficient to be multiplied by the compression force. For clays it is given as a specific value in pounds per square foot to be multiplied by the contact area.

Whenever possible, stress limits should be established as the result of a thorough investigation and the recommendations of a qualified soils engineer. Most building codes allow for the use of so-called *presumptive* values for design. These are average values, on the conservative side usually, that may be used for soils identified by groupings used by the codes. Table 9.1 is a reproduction of such a source, taken from the 1988 edition of the *Uniform Building Code* (Ref. 4).

For very ordinary situations, building codes or code enforcing agencies sometimes permit construction without submission of detailed engineering design. Table 9.2 is a reproduction from the 1988 edition of the *Uniform Building Code,* yielding data for footings for simple stud bearing wall construction of light wood frame. In a similar manner, the data displayed in Fig. 9.1 yield information on various situations with similar construction.

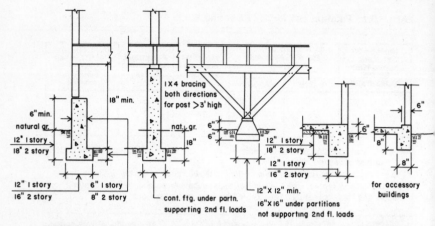

FIGURE 9.1 Typical recommended details for foundations for light wood frame construction. Adapted from an information sheet for type V construction issued by the Department of Building and Safety of the City of Los Angeles.

9.4 SHALLOW BEARING FOUNDATIONS

The most common foundation consists of simple pads of concrete placed immediately beneath the building. Because most buildings make a relatively shallow penetration into the ground, these pads—called *footings*—are generally classified as *shallow bearing foundations*. This is opposed to the use of piles or excavated piers (caissons) which extend some distance below the building and are called *deep foundations*.

The two common forms of footings are the wall footing and the isolated column footing. Wall footings occur in strip form, usually placed symmetrically beneath the supported wall. Column footings are most often simple square pads supporting a single column. When columns are very close together or at the very edge of the building site, special footings carrying more than a single column may be used. Some of these types of footings are discussed in Sec. 9.7.

9.5   WALL FOOTINGS

Wall footings consist of concrete strips placed under walls. The most common type of wall footing is that shown in Fig. 9.2, consisting of a strip with a rectangular cross section placed in a symmetrical position with respect to the wall and projecting an equal distance as a cantilever from both faces of the wall. For soil stress the critical dimension of the footing is the width of the footing bottom measured perpendicular to the wall face.

In most situations, the wall footing is utilized as a platform upon which the wall is constructed. Thus a minimum width for the footing is established by the wall thickness, the footing usually being made somewhat wider than the wall. With a concrete wall this additional width is used to support the wall forms while the concrete is poured. For masonry walls this added width assures an adequate base for the mortar bed for the first course of the masonry units. The exact additional width required for these purposes is a matter of judgment. For support of concrete forms, it is usually desirable to have at least a 3-in. projection; for masonry, the usual minimum is 2 in.

With relatively lightly loaded walls, the minimum width required for platform considerations may be more than adequate in terms of the allowable bearing stress on the soil. If this is the case, the short projection of the footing from the wall face will produce relatively insignificant transverse bending and shear stresses, per-

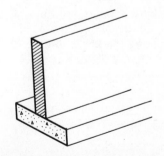

FIGURE 9.2   Continuous wall footing.

mitting a minimal thickness for the footing and the omission of transverse reinforcing. Most designers prefer, however, to provide some continuous reinforcing in the long direction of the footing, even when none is used in the transverse direction. The purpose is to reduce shrinkage cracking and also to give some enhanced beamlike capabilities for spanning over soft spots in the supporting soil.

As the wall load increases, the increased width of the footing required to control soil stress eventually produces significant transverse bending and shear in the footing. At some point this determines the required thickness for the footing and for required reinforcing in the transverse direction. If the footing is not reinforced in the transverse direction, the controlling stress is usually the transverse tensile bending stress in the concrete. If the footing has transverse reinforcing, the controlling concrete stress is usually the shear stress.

The following example illustrates the procedure for design of a wall footing with transverse reinforcing.

Example. Using concrete with $f'_c = 2$ ksi and grade 40 reinforcing, design a wall footing for the following data: wall thickness = 6 in.; load on footing = 8750 lb/ft of wall length; maximum allowable soil pressure = 2000 psf.

Solution: For the wall footing, the only concrete stress of concern is that of shear. Compression stress in flexure is seldom critical due to design for shear and the desire for minimal transverse reinforcing. The usual design procedure consists of making a guess for the footing thickness and determining conditions to verify the guess. Code restrictions establish a minimum thickness of 10 in. Try

$$h = 12 \text{ in.}$$

Then

$$\text{footing weight} = 150 \text{ psf}$$

$$\text{usable soil pressure} = 2000 - 150 = 1850 \text{ psf}$$

$$\text{required width} = \frac{8750}{1850} = 4.73 \text{ ft. or } 56.8 \text{ in.}$$

Try

$$\text{width} = 57 \text{ in. or 4 ft 9 in.}$$

Then

$$\text{design soil pressure} = \frac{8750}{4.75} = 1842 \text{ psf}$$

With 3-in. cover and a No. 6 bar (a guess), the effective depth will be 8.625 in., say 8.6 in. approximately.

The ACI Code requires that shear be investigated as a beam shear condition, with the critical section at a distance d (effective depth) from the face of the wall. This condition is shown in Fig. 9.3a. This is reasonably valid when the cantilever distance is larger than the footing thickness by a significant amount, but is questionable for short cantilevers. In fact, the code recommends that this shortened span not be used for brackets and short cantilevers. Therefore, we recommend that the critical section for shear be taken at the face of the wall unless the cantilever exceeds three times the overall footing thickness. However, if the latter assumption is made (short cantilever analysis), it is reasonable to use the full thickness of the footing for the stress computation rather than the effective depth of the cross section. Both cases are shown in Fig. 9.3, and we will show the computations for both.

Case 1. Shear at the d distance from the wall (see Fig. 9.3a):

$$V = 1842 \left(\frac{16.9}{12}\right) = 2594 \text{ lb}$$

Stress:

$$v_c = \frac{V}{bd} = \frac{2594}{12(8.6)} = 25 \text{ psi}$$

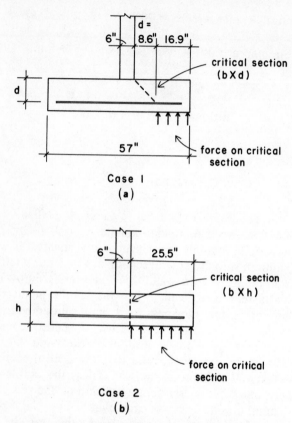

FIGURE 9.3 Shear considerations for the wall footing.

Case 2. Shear at the wall face (see Fig. 9.3*b*):

$$V = 1842 \frac{25.5}{12} = 3914 \text{ lb}$$

Stress:

$$v_c = \frac{V}{bh} = \frac{3914}{12(12)} = 27 \text{ psi}$$

Both of these are well below the allowable stress of $1.1 \sqrt{f'_c}$, which is the same as in the previous example: 49 psi. It is possible, therefore, to reduce the footing thickness if the shear stress is considered to be an important criterion. However, as has been discussed previously, reduction of cost in construction is usually obtained by minimizing the amount of reinforcing, and any reduction in the footing thickness will shorten the moment arm for the tension reinforcing, requiring an increase in steel area. It therefore becomes a matter of judgment about the ideal value for the footing thickness.

If we reduce the footing thickness to 11 in., a second try would proceed as follows:

$$\text{new footing weight} = \frac{11}{12} (150) = 137.5, \text{ say } 138 \text{ lb/ft}^2$$

$$\text{usable soil pressure } (p) = 2000 - 138 = 1862 \text{ lb/ft}^2$$

$$\text{required width } (w) = \frac{8750}{1862} = 4.70 \text{ ft or } 56.4 \text{ in.}$$

which does not change the footing width or design soil pressure.

$$\text{new } d = h - 3 - \frac{D}{2} = 11 - 3.375 = 7.625, \text{ say } 7.6 \text{ in.}$$

For the case 2 shear stress, the shear force is the same as for the first try, and the new shear stress is

$$v_c = \frac{V}{bh} = \frac{3914}{12(11)} = 30 \text{ psi}$$

For the case 1 shear stress, the shear section is now an inch closer to the wall and the shear force becomes

$$V = 1842 \left(\frac{17.9}{12}\right) = 2748 \text{ lb}$$

$$v_c = \frac{V}{bd} = \frac{2748}{12(7.6)} = 30 \text{ psi}$$

The bending moment to be used for concrete stress and determination of the steel area is

$$M = 3914 \left(\frac{25.5}{2}\right) = 49{,}903 \text{ lb-in.}$$

and the required steel area per foot of wall length is

$$A_s = \frac{M}{f_s j d} = \frac{49{,}903}{20(0.9)(7.6)} = 0.365 \text{ in.}^2$$

Since the steel area requirement has been determined in the same manner as for a slab, Table 6.4 may be used to select the bars and their spacing. The following should be considered in making the selection.

1. Maximum recommended spacing is 18 in.
2. Minimum recommended spacing is 6 in. to minimize the number of bars and allow for easy placing of the concrete during construction.
3. For proper development of the bars smaller bar sizes are usually preferable.

Table 9.3 presents a summary of the possible alternatives for reinforcing in the transverse direction, as determined from the

TABLE 9.3 Selection of Reinforcing for Example

Bar Size	Area of Bar (in.^{2})	Area Required for Flexure (in.^{2})	Spacing Required (in.)	Selected Spacing (in.)
3	0.11	0.365	3.6	3.5
4	0.20	0.365	6.6	6.5
5	0.31	0.365	10.2	10
6	0.44	0.365	14.5	14.5
7	0.60	0.365	19.7	19.5

data in Table 6.4. Our preference would be for the No. 5 bars at 10 in. center to center. Reference to Table 6.6 will show that development is more than adequate for these bars. (*Note:* Data are not given in Table 6.6 for $f'_c = 2000$ psi; however, the table shows only 12 in. required for $f'_c = 3000$ psi, whereas over 23 in. is available with our footing.)

Whether transverse reinforcing is provided or not, we recommend a minimum reinforcing for shrinkage stresses in the long direction of the footing consisting of 0.0015 times the gross area of the cross section. Thus

$$A_s = 0.0015 \times 11 \times 57 = 0.94 \text{ in.}^2$$

This area can be supplied by using three No. 5 bars with a total area of 0.93 in.^{2}.

Table 9.4 gives values for wall footings for four different soil pressures. Table data were derived using the procedures illustrated in the example. Figure 9.4 shows the dimensions referred to in the table.

PROBLEM 9.5.A. Using concrete with $f'_c = 2$ ksi [13.8 MPa] and grade 40 bars with $f_y = 40$ ksi [276 MPa] and $f_s = 20$ ksi [138 MPa], design a wall footing for the following data: wall thickness = 10 in. [254 mm]; load on footing = 12,000 lb/ft [175 kN/m]; maximum soil pressure = 2000 psf [96 kN/m²].

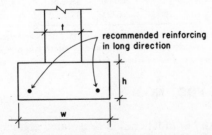

FIGURE 9.4 Reference figure for Table 9.4.

TABLE 9.4 Allowable Loads on Wall Footings (see Fig. 9.4)

Maximum Soil Pressure (lb/ft²)	Minimum Wall Thickness, t (in.)		Allowable Load on Footing[a] (lb/ft)	Footing Dimensions (in.)		Reinforcing	
	Concrete	Masonry		h	w	Long Direction	Short Direction
1000	4	8	2625	10	36	3 No. 4	No. 3 at 16
	4	8	3062	10	42	2 No. 5	No. 3 at 12
	6	12	3500	10	48	4 No. 4	No. 4 at 16
	6	12	3938	10	54	3 No. 5	No. 4 at 13
	6	12	4375	10	60	3 No. 5	No. 4 at 10
	6	12	4812	10	66	5 No. 4	No. 5 at 13
	6	12	5250	10	72	4 No. 5	No. 5 at 11
1500	4	8	4125	10	36	3 No. 4	No. 3 at 10
	4	8	4812	10	42	2 No. 5	no. 4 at 13
	6	12	5500	10	48	4 No. 4	No. 4 at 11
	6	12	6131	11	54	3 No. 5	No. 5 at 15
	6	12	6812	11	60	5 No. 4	No. 5 at 12
	6	12	7425	12	66	4 No. 5	No. 5 at 11
	8	16	8100	12	72	5 No. 5	No. 5 at 10
2000	4	8	5625	10	36	3 No. 4	No. 4 at 14
	6	12	6562	10	42	2 No. 5	No. 4 at 11
	6	12	7500	10	48	4 No. 4	No. 5 at 12
	6	12	8381	11	54	3 No. 5	No. 5 at 11
	6	12	9520	12	60	4 No. 5	No. 5 at 10
	8	16	10106	13	66	4 No. 5	No. 5 at 9
	8	16	10875	15	72	6 No. 5	No. 5 at 9
3000	6	12	8625	10	36	3 No. 4	No. 4 at 10
	6	12	10019	11	42	4 No. 4	No. 5 at 13
	6	12	11400	12	48	3 No. 5	No. 5 at 10
	6	12	12712	14	54	6 No. 4	No. 5 at 10
	8	16	14062	15	60	5 No. 5	No. 5 at 9
	8	16	15400	16	66	5 No. 5	No. 6 at 12
	8	16	16725	17	72	6 No. 5	No. 6 at 10

[a] Allowable loads do not include the weight of the footing, which has been deducted from the total bearing capacity. Criteria: $f'_c = 2000$ psi, grade 40 reinforcing, $v_c = 1.1 \sqrt{f'_c}$.

9.6 COLUMN FOOTINGS

The great majority of independent or isolated column footings are square in plan, with reinforcing consisting of two sets of bars at right angles to each other. This is known as two-way reinforcement. The column may be placed directly on the footing block, or it may be supported by a pedestal. A pedestal, or pier, is a short, wide compression block that serves to reduce the punching effect on the footing. For steel columns a pier may also serve to raise the bottom of the steel column above ground level.

The design of a column footing is usually based on the following considerations:

Maximum Soil Pressure. The sum of the superimposed load on the footing and the weight of the footing must not exceed the limit for bearing pressure on the supporting material. The required total plan area of the footing is determined on this basis.

Control of Settlement. Where buildings rest on highly compressible soil, it may be necessary to select footing areas that assure a uniform settlement of all the building columns rather than to strive for a maximum use of the allowable soil pressure.

Size of the Column. The larger the column, the less will be the shear, flexural, and bond stresses in the footing, since these are developed by the cantilever effect of the footing projection beyond the edges of the column.

Shear Stress Limit for the Concrete. For square-plan footings this is usually the only critical stress condition for the concrete. In order to reduce the required amount of reinforcing, the footing depth is usually established well above that required by the flexural stress limit for the concrete.

Flexural Stress and Development Length Limits for the Bars. These are considered on the basis of the moment developed in the cantilevered footing at the face of the column.

Footing Thickness for Development of Column Reinforcing. When a footing supports a reinforced concrete column, the compressive force in the column bars must be transferred to the footing by bond stress—called *doweling* of the bars. The thickness of the footing must be sufficient for the necessary development length of the column bars (see Sec. 6.11).

The following example illustrates the design process for a simple, square column footing.

Example. A 16-in. [406-mm] square concrete column exerts a load of 240 kips [1068 kN] on a square column footing. Determine the footing dimensions and the necessary reinforcing using the following data: f'_c = 3 ksi [20.7 MPa], grade 40 bars with f_y = 40 ksi [276 MPa] and f_s = 20 ksi [138 MPa], maximum permissible soil pressure = 4000 psf [192 kPa].

Solution: The first decision to be made is that of the height, or thickness, of the footing. This has to be a raw first guess unless the dimensions of similar footings are known. In practice this knowledge is generally available from previous design work or from handbook tables. In lieu of this, a reasonable guess is made, the design work is performed, and an adjustment is made if the assumed thickness proves inadequate. We will assume a footing thickness of 20 in. [508 mm] for a first try for this example.

The footing thickness establishes the weight of the footing on a per-square-ft basis. This weight is then subtracted from the maximum permissible soil pressure, and the net value is then usable for the superimposed load on the footing. Thus

$$\text{footing weight} = \frac{20}{12}\,(150 \text{ psf}) = 250 \text{ psf } [12 \text{ kPa}]$$

$$\text{net usable soil pressure} = 4000 - 250 = 3750 \text{ psf } [180 \text{ kPa}]$$

$$\text{required footing plan area} = \frac{240{,}000}{3750} = 64 \text{ ft}^2 \ [5.93 \text{ m}^2]$$

$$\text{length of the side of the square footing} = L = \sqrt{64}$$
$$= 8 \text{ ft } [2.44 \text{ m}]$$

Two shear stress situations must be considered for the concrete. The first occurs as ordinary beam shear in the cantilevered portion and is computed at a critical section at a distance d (effective depth of the beam) from the face of the column, as shown in

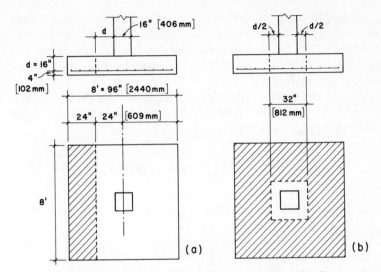

FIGURE 9.5 Shear considerations for the column footing.

Fig. 9.5*a*. The shear stress at this section is computed in the same manner as for a beam, as discussed in Sec. 6.9, and the stress limit is $v_c = 1.1 \sqrt{f'_c}$. The second shear stress condition is that of peripheral shear, or so-called "punching" shear, and is investigated at a circumferential section around the column at a distance of $d/2$ from the column face, as shown in Fig. 9.5*b*. For this condition the allowable stress is $v_c = 2.0 \sqrt{f'_c}$.

With two-way reinforcing, it is necessary to place the bars in one direction on top of the bars in the other direction. Thus, although the footing is supposed to be the same in both directions, there are actually two different d distances—one for each layer of bars. It is common practice to use the average of these two distances for the design value of d; that is, d = the footing thickness less the sum of the concrete cover and the bar diameter. With the bar diameter as yet undetermined, we will assume an approximate d of the footing thickness less 4 in. [102 mm] (a concrete cover of 3 in. plus a No. 8 bar). For example this becomes

$$d = t - 4 = 20 - 4 = 16 \text{ in. [406 mm]}$$

It should be noted that it is the *net* soil pressure that causes stresses in the footing, since there will be no bending or shear in the footing when it rests alone on the soil. We thus use the net soil pressure of 3750 psf [180 kPa] to determine the shear and bending effects for the footing.

For the beam shear investigation, we determine the shear force generated by the net soil pressure acting on the shaded portion of the footing plan area shown in Fig. 9.5*a*. Thus

$$V = 3750 \times 8 \times \frac{24}{12} = 60,000 \text{ lb } [267.5 \text{ kN}]$$

and using the formula for shear stress in a beam (Sec. 6.9),

$$v = \frac{V}{bd} = \frac{60,000}{96 \times 16} = 39.1 \text{ psi } [0.270 \text{ MPa}]$$

which is compared to the allowable stress of

$$v_c = 1.1 \sqrt{f_c'} = 1.1 \sqrt{3000} = 60 \text{ psi } [0.414 \text{ MPa}]$$

indicating that this condition is not critical.

For the peripheral shear investigation, we determine the shear force generated by the net soil pressure acting on the shaded portion of the footing area shown in Fig. 9.5*b*. Thus

$$V = 3750 \left[8^2 - \left(\frac{32}{12} \right)^2 \right] = 213,333 \text{ lb } [953 \text{ kN}]$$

Shear stress for this case is determined with the same formula as for beam shear, with the dimension *b* being the total peripheral circumference. Thus

$$v = \frac{V}{bd} = \frac{213,333}{(4 \times 32) \times 16} = 104.2 \text{ psi } [0.723 \text{ MPa}]$$

which is compared to the allowable stress of

$$v_c = 2 \sqrt{f_c'} = 2 \sqrt{3000} = 109.5 \text{ psi } [0.755 \text{ MPa}]$$

This computation indicates that the peripheral shear stress is not critical, but since the actual stress is quite close to the limit, the assumed thickness of 20 in. is probably the least full-inch value that can be used. Flexural stress in the concrete should also be considered, although it is seldom critical for a square footing. One way to verify this is to compute the balanced moment capacity of the section with $b = 96$ in. and $d = 16$ in. Using the factor for a balanced section from Table 6.1, we find

$$M_R = Rbd^2 = 0.226(96)(16)^2 = 5554 \text{ kip-in or } 463 \text{ kip-ft}$$

which may be compared with the actual moment computed in the next step.

For the reinforcing we consider the stresses developed at a section at the edge of the column, as shown in Fig. 9.6. The cantilever moment for the 40-in. [1016-mm] projection of the foot-

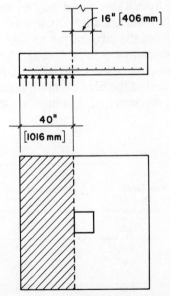

FIGURE 9.6 Bending and bar development considerations for the column footing.

ing beyond the column is

$$M = 3750 \times 8 \times \frac{40}{12} \times \frac{1}{2}\left(\frac{40}{12}\right) = 166,667 \text{ lb-ft [227 kN-m]}$$

Using the formula for required steel area in a beam, with a conservative guess of 0.9 for j, we find (see Sec. 6.4)

$$A_s = \frac{M}{f_s jd} = \frac{166,667 \times 12}{20 \times 0.9 \times 16 \times 10^3} = 6.95 \text{ in.}^2 \text{ [4502 mm}^2\text{]}$$

This requirement may be met by various combinations of bars, such as those in Table 9.5. Data for consideration of the development length and the center-to-center bar spacing are also given in the table. The flexural stress in the bars must be developed by the embedment length equal to the projection of the bars beyond the column edge, as discussed in Sec. 6.11. With a minimum of 2 in. [51 mm] of concrete cover at the edge of the footing, this length is 38 in. [965 mm]. The required development lengths indicated in the table are taken from Table 6.6; it may be noted that all of the combinations in the table are adequate in this regard.

If the distance from the edge of the footing to the first bar at each side is approximately 3 in. [76 mm], the center-to-center distance for the two outside bars will be $96 - 2(3) = 90$ in. [2286 mm], and with the rest of the bars evenly spaced, the spacing will be 90 divided by the number of total bars less one. This value is

TABLE 9.5 Reinforcing Alternatives for the Column Footing

Number and Size of Bars	Area of Steel Provided		Required Development Length[a]		Center-to-Center Spacing	
	in.²	mm²	in.	mm	in.	mm
12 No. 7	7.20	4645	18	457	8.2	208
9 No. 8	7.11	4687	23	584	11.3	286
7 No. 9	7.00	4516	29	737	15	381
6 No. 10	7.62	4916	37	940	18	458

[a] From Table 6.6; values for "other bars," $f_y = 40$ ksi, $f_c' = 3$ ksi.

ABLE 9.6 Square Column Footings

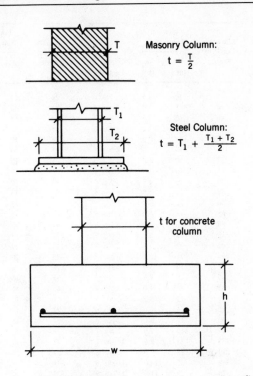

Masonry Column:
$$t = \frac{T}{2}$$

Steel Column:
$$t = T_1 + \frac{T_1 + T_2}{2}$$

t for concrete column

		$f'_c = 2000$ psi				$f'_c = 3000$ psi			
			Footing Dimensions				Footing Dimensions		
Maximum Soil Pressure (lb/ft²)	Minimum Column Width t (in.)	Allowable Load[a] on Footing (k)	h (in.)	w (ft)	Reinforcing Each Way	Allowable Load[a] on Footing (k)	h (in.)	w (ft)	Reinforcing Each Way
1000	8	7.9	10	3.0	2 No. 3	7.9	10	3.0	2 No. 3
	8	10.7	10	3.5	3 No. 3	10.7	10	3.5	3 No. 3
	8	14.0	10	4.0	3 No. 4	14.0	10	4.0	3 No. 4
	8	17.7	10	4.5	4 No. 4	17.7	10	4.5	4 No. 4
	8	22	10	5.0	4 No. 5	22	10	5.0	4 No. 5
	8	31	10	6.0	5 No. 6	31	10	6.0	5 No. 6
	8	42	12	7.0	6 No. 6	42	11	7.0	7 No. 6
1500	8	12.4	10	3.0	3 No. 3	12.4	10	3.0	3 No. 3
	8	16.8	10	3.5	3 No. 4	16.8	10	3.5	3 No. 4
	8	22	10	4.0	4 No. 4	22	10	4.0	4 No. 4

TABLE 9.6 (Continued)

Maximum Soil Pressure (lb/ft²)	Minimum Column Width t (in.)	f'_c = 2000 psi				f'_c = 3000 psi			
		Allowable Load[a] on Footing (k)	h (in.)	w (ft)	Reinforcing Each Way	Allowable Load[a] on Footing (k)	h (in.)	w (ft)	Reinforci Each Wa
	8	28	10	4.5	4 No. 5	28	10	4.5	4 No. 5
	8	34	11	5.0	5 No. 5	34	10	5.0	6 No. 5
	8	48	12	6.0	6 No. 6	49	11	6.0	6 No. 6
	8	65	14	7.0	7 No. 6	65	13	7.0	6 No. 7
	8	83	16	8.0	7 No. 7	84	15	8.0	7 No. 7
	8	103	18	9.0	8 No. 7	105	16	9.0	10 No. 7
2000	8	17	10	3.0	4 No. 3	17	10	3.0	4 No. 3
	8	23	10	3.5	4 No. 4	23	10	3.5	4 No. 4
	8	30	10	4.0	6 No. 4	30	10	4.0	6 No. 4
	8	37	11	4.5	5 No. 5	38	10	4.5	6 No. 5
	8	46	12	5.0	6 No. 5	46	11	5.0	5 No. 6
	8	65	14	6.0	6 No. 6	66	13	6.0	7 No. 6
	8	88	16	7.0	8 No. 6	89	15	7.0	7 No. 7
	8	113	18	8.0	8 No. 7	114	17	8.0	9 No. 7
	8	142	20	9.0	8 No. 8	143	19	9.0	8 No. 8
	10	174	21	10.0	9 No. 8	175	20	10.0	10 No. 8
3000	8	26	10	3.0	3 No. 4	26	10	3.0	3 No. 4
	8	35	10	3.5	4 No. 5	35	10	3.5	4 No. 5
	8	45	12	4.0	4 No. 5	46	11	4.0	5 No. 5
	8	57	13	4.5	6 No. 5	57	12	4.5	6 No. 5
	8	70	14	5.0	5 No. 6	71	13	5.0	6 No. 6
	8	100	17	6.0	7 No. 6	101	15	6.0	8 No. 6
	10	135	19	7.0	7 No. 7	136	18	7.0	8 No. 7
	10	175	21	8.0	10 No. 7	177	19	8.0	8 No. 8
	12	219	23	9.0	9 No. 8	221	21	9.0	10 No. 8
	12	269	25	10.0	11 No. 8	271	23	10.0	10 No. 9
	12	320	28	11.0	11 No. 9	323	26	11.0	12 No. 9
	14	378	30	12.0	12 No. 9	381	28	12.0	11 No. 1
4000	8	35	10	3.0	4 No. 4	35	10	3.0	4 No. 4
	8	47	12	3.5	4 No. 5	47	11	3.5	4 No. 5
	8	61	13	4.0	5 No. 5	61	12	4.0	6 No. 5
	8	77	15	4.5	5 No. 6	77	13	4.5	6 No. 6
	8	95	16	5.0	6 No. 6	95	15	5.0	6 No. 6
	8	135	19	6.0	8 No. 6	136	18	6.0	7 No. 7
	10	182	22	7.0	8 No. 7	184	20	7.0	9 No. 7
	10	237	24	8.0	9 No. 8	238	22	8.0	9 No. 8
	12	297	26	9.0	10 No. 8	299	24	9.0	9 No. 9
	12	364	29	10.0	13 No. 8	366	27	10.0	11 No. 9
	14	435	32	11.0	12 No. 9	440	29	11.0	11 No. 1(
	14	515	34	12.0	14 No. 9	520	31	12.0	13 No. 1(
	16	600	36	13.0	17 No. 9	606	33	13.0	15 No. 1(
	16	688	39	14.0	15 No. 10	696	36	14.0	14 No. 1
	18	784	41	15.0	17 No. 10	793	38	15.0	16 No. 1

[a] Allowable loads do not include the weight of the footing, which has been deducted from the total bearing capacity. Criteria f'_c = 20 ksi, v_c = 1.1 $\sqrt{f'_c}$ for beam shear, v_c = 2 $\sqrt{f'_c}$ for peripheral shear.

shown in the table for each set of bars. The maximum permitted spacing is 18 in. [457 mm], and the minimum should be a distance that is adequate to permit good flow of the wet concrete between the two-way grid of bars—say 4 in. [102 mm] or more.

All of the bar combinations in Table 9.5 are adequate for the footing. Many designers prefer to use the largest possible bar, as this reduces the number of bars that must be handled and supported during construction. On this basis, the footing will be the following:

8 ft square by 20 in. thick with 6 No. 10 bars each way.

For ordinary situations we often design square column footings by using data from tables in various references. Even where special circumstances make it necessary to perform the type of design illustrated in the example, such tables will assist in making a first guess for the footing dimensions.

Table 9.6 gives the allowable superimposed load for a range of footings and soil pressures. This material has been adapted from a more extensive table in *Simplified Design of Building Foundations* (Ref. 13). Designs are given for concrete strengths of 2000 and 3000 psi. The low strength of 2000 psi is sometimes used for small buildings, since many building codes permit the omission of testing the concrete if this value is used for design.

Problem 9.6.A. Design a square footing for a 14-in. [356-mm] square concrete column with a load of 219 kips [974 kN]. The maximum permissible soil pressure is 3000 psf [144 kPa]. Use concrete with $f'_c = 3$ ksi [20.7 MPa] and reinforcing of grade 40 bars with $f_y = 40$ ksi [276 MPa] and $f_s = 20$ ksi [138 MPa].

9.7 SPECIAL COLUMN FOOTINGS

Situations commonly occur in which a simple, square footing may not be indicated for a column. The following are some special forms for column footings for frequently encountered situations.

Rectangular Footing. When soil design pressures are low, or a column must be placed close to some other construction, it may be necessary to use a footing that is oblong in plan, rather than square, a form referred to as a rectangular footing. Design is performed essentially as for a square footing, except that there are special requirements for placing of the reinforcement in the short direction.

Combined Footing. When two or more columns are placed close together in the building plan, a single footing is sometimes

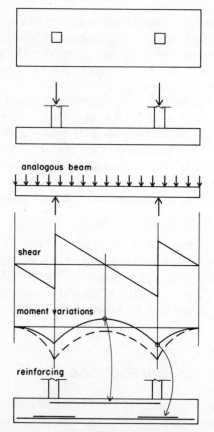

FIGURE 9.7 Actions of a symmetrical combined footing for two columns.

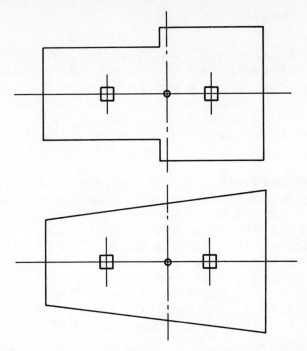

FIGURE 9.8 Plan variations for an unsymmetrical combined footing.

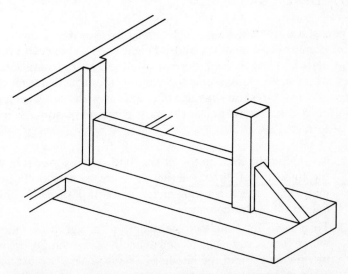

FIGURE 9.9 Use of a cantilever footing with a stiffening element.

used, called a combined footing. An example of such a footing for two equally loaded columns is shown in Fig. 9.7. In this case a simple oblong footing is placed symmetrically beneath the columns and is designed as a double cantilevered beam. If the columns are not equally loaded, the footing may be shifted to have its plan centroid coincide with that of the column loads, or some other form, such as those shown in Fig. 9.8, may be used.

Cantilever Footing.   A common situation that occurs when building on tight urban sites is one where the edge of the building is placed very close to the property line. If a column occurs at the edge of the building, a conventional footing would likely extend a considerable distance beyond the edge of the building. In such a case, one solution is to use a cantilever footing, also called a strap footing, consisting of a combined footing supporting the exterior column and an adjacent interior column. Such a footing is shown in Fig. 9.9, with a stiffening stem wall to form a T-beam action for the major bending that occurs midway between the columns.

9.8 PEDESTALS

A pedestal (also called a pier) is defined by the ACI Codes as a short compression member whose height does not exceed three times its width. Pedestals are frequently used as transitional elements between columns and the bearing footings that support them. Figure 9.10 shows the use of pedestals with both steel and reinforced concrete columns. The most common reasons for use of pedestals are:

1. To spread the load on top of the footing. This may relieve the intensity of direct bearing pressure on the footing or may simply permit a thinner footing with less reinforcing due to the wider column.
2. To permit the column to terminate at a higher elevation where footings must be placed at depths considerably below the lowest parts of the building. This is generally most significant for steel columns.

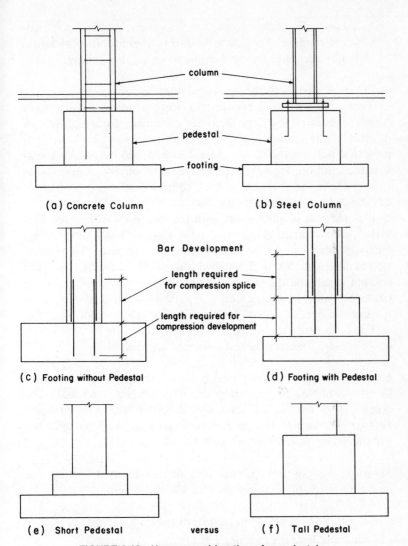

FIGURE 9.10 Usage considerations for pedestals.

3. To provide for the required development length of reinforcing in reinforced concrete columns, where footing thickness is not adequate for development within the footing.

Figure 9.10c illustrates the third situation described. Referring to Table 6.8, we may observe that a considerable development length is required for large diameter bars made from high grades of steel. If the minimum required footing does not have a thickness that permits this development, a pedestal may offer a reasonable solution. However, there are many other considerations to be made in the decision, and the column reinforcing problem is not the only factor in this situation.

If a pedestal is quite short with respect to its width (see Fig. 9.10e), it may function essentially the same as a column footing, with significant values for shear and bending stresses. This condition is likely to occur if the pedestal width exceeds twice the column width and the pedestal height is less than one-half of the pedestal width. In such cases, the pedestal must be designed by the same procedures used for an ordinary column footing.

The following example illustrates the procedure for the design of a pedestal for a reinforced concrete column.

Example. A 16-in. square tied column with f'_c of 4 ksi is reinforced with No. 10 bars of grade 60 steel (F_y = 60 ksi). The column axial load is 200 kips, and the allowable maximum soil pressure is 4000 psf. Design a footing and a pedestal, using f'_c = 3 ksi and grade 40 reinforcing with f_y = 40 ksi.

Solution: For an approximate idea of the required footing, we may refer to Table 9.6 and observe the following.

8-ft square footing, 22 in. thick, nine No. 8 each way.

Allowable load on footing: 238 kips.

Designed for column width of 10 in.

From Table 6.8, for No. 10 bar, grade 60, we observe the following:

$$f'_c = 3 \text{ ksi}, \qquad l_d = 27.8 \text{ in.}$$

From these observations, we may conclude that:

1. The minimum required footing for the 16-in. column with 200-kip load will be slightly smaller than that taken from the table. Thus it will not be adequate for development of the column bars.
2. If a pedestal is used, it must be at least 28 in. high to develop the column bars.
3. With a pedestal slightly wider than the column, the footing thickness may be additionally reduced, if shear stress is the critical design factor for the footing thickness.

One option in this case is to simply forget about a pedestal and increase the footing thickness to that required for development of the column bars. This means an increase from around 20 in. up to 31 in., giving the necessary 28 in. of development plus 3 in. of cover. Let us therefore consider the possibility of a footing that is 7.5 ft square and 31 in. thick. Then

$$\text{design soil pressure} = \frac{200{,}000}{(7.5)^2} = 3556 \text{ psf}$$

Adding the weight of the footing to this, the total soil pressure becomes

$$3556 + \frac{31}{12}(150) = 3944 \text{ psf}$$

which is less than the allowable of 4000 psf, so the footing width is adequate.

Shear stress is obviously not a critical concern, so we proceed to determine the required reinforcing. Figure 9.11a indicates the basis for determining the cantilever moment, and we thus compute the following.

$$M = 3.556(7.5)\left(\frac{37}{12}\right)^2(12) = 127 \text{ kip-ft}$$

$$A_s = \frac{M}{f_s jd} = \frac{127(12)}{20(0.9)(27)}$$
$$= 3.14 \text{ in.}^2$$

Try six No. 7 bars:

$$A_s = 3.6 \text{ in.}^2$$

Table 6.6 will indicate that the 37-in. projection is adequate for development of the No. 7 bars. It may be noted that this is considerably less reinforcing than that given for the footing taken from Table 9.6.

If it is desired to use a pedestal, we consider the use of the one shown in Fig. 9.11b. The 28-in. height shown is the minimum

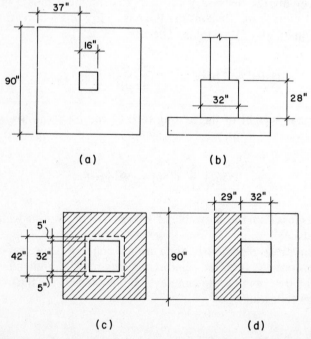

(a)

(b)

(c)

(d)

FIGURE 9.11 Form and loading considerations for the footing and pedestal.

established previously for the development of the column bars. The height could be increased to as much as 96 in. (three times the width), if it is desired for other reasons. One such reason may be the presence of a better soil for bearing at a lower elevation.

A potential concern is that for the direct bearing of the column on the pedestal. If the pedestal is designed as an unreinforced member, the ACI Code permits a maximum bearing stress of

$$f_p = 0.3 f'_c \sqrt{\frac{A_2}{A_1}}$$

where A_1 is the actual bearing area (in our case the 16-in. square column area) and A_2 is the area of the pedestal cross section. The maximum usable value for $\sqrt{A_2/A_1}$ is 2.

In our example it will be found that the allowable stress thus determined is more than twice the value of the direct bearing, even if the latter is computed ignoring the portion of the load transferred by development of the column reinforcing. The only times that this condition is likely to be critical is when a pedestal with very low f'_c supports a column with very high f'_c and the pedestal width is only slightly greater than the column width. When the pedestal supports a steel column, however, this condition may be the basis for establishing the width of the pedestal.

Another consideration for bearing stress is that of the pedestal on the footing. In this case, using the same criteria described previously, the maximum allowable bearing stress will be either $0.3 f'_c$ for the pedestal or $0.6 f'_c$ (with the maximum value of $\sqrt{A_2/A_1}$) for the footing. This is also not critical for our example.

If the pedestal height exceeds its width, we recommend the use of a minimum of column reinforcing of not less than $A_s = 0.005 A_g$. This should be installed with at least four bars, one in each corner, and a set of loop ties, just as with an ordinary tied column. For our short pedestal, this is of questionable necessity.

With the wide pedestal, the footing thickness can be reduced considerably, if the minimum thickness for shear stress is desired. The basis for consideration of peripheral shear stress is shown in Fig. 9.11c, and the computations are as follows. For the example, we have assumed a footing thickness of 14 in. with an effective depth of 10 in.

$$\text{weight of pedestal} = \frac{32(32)(28)}{1728}(150) = 2489 \text{ lb}$$

$$\text{design soil pressure} = \frac{202,489}{(7.5)^2} = 3600 \text{ psf}$$

$$V = 36\left\{(7.5)^2 - \left(\frac{42}{12}\right)^2\right\} = 158.4 \text{ kips}$$

$$v = \frac{V}{bd} = \frac{158,400}{4(42)(10)} = 94.3 \text{ psi} < 110 \text{ psi}$$

The basis for consideration of the reinforcing is shown in Fig. 9.11d, and the computations are as follows.

$$M = 3.6(7.5)\left(\frac{29}{12}\right)^2\left(\frac{1}{2}\right) = 78.8 \text{ kip-ft}$$

$$A_s = \frac{M}{f_s j d} = \frac{78.8(12)}{20(0.9)(10)} = 5.25 \text{ in.}^2$$

This could be supplied by using seven No. 8 bars each way. It may be noted that this is considerably more reinforcing than that required for the thickened footing without the pedestal. Cost savings affected by the pedestal may thus be questionable, and its use may depend on its need for other purposes.

PROBLEM 9.8.A. An 18-in square tied column with $f'_c = 4$ ksi [27.6 MPa] is reinforced with No. 11 bars of grade 60 steel with $f_y = 60$ ksi [414 MPa]. The column axial load is 260k [1156 kN] and the allowable soil pressure is 3000 psf [134 kN/m^2]. Using $f'_c = 3$ ksi [20.7 MPa] and grade 40 bars with $f_y = 40$ ksi [276 MPa] and $f_s = 20$ ksi [138 MPa], design the following: (1) a footing without a pedestal and (2) a footing with a pedestal.

9.9 PAVING SLABS

Sidewalks, driveways, and basement floors are typically produced by depositing a relatively thin coating of concrete directly on the ground surface. While the basic construction process is simple, a number of factors must be considered in developing details and specifications for a paving slab.

Thickness of the Slab

Pavings vary in thickness from a few inches (for residential basement floors) to several feet (for airport landing strips). Although more strength is implied by a thicker slab, thickness alone does not guarantee a strong pavement. Of equal concern is the reinforcement provided and the character of the subbase on which the concrete is poured. The minimum slab thickness commonly used in building floor slabs is $3\frac{1}{2}$ in. This relates specifically to the actual dimension of a nominal wood 2×4, and simplifies forming the edges of a slab pour. Following the same logic, the next-size jump would be to a $5\frac{1}{2}$-in.-thickness, which is the dimension of a nominal 2×6.

The $3\frac{1}{2}$-in.-thick slab is usually considered adequate for interior floors not subjected to wheel loadings or other heavy structural demand. At this thickness, usually provided with very minimal reinforcing, the slab has relatively low resistance to bending and shear effects of concentrated loads. Thus walls, columns, and heavy items of equipment should be provided with separate footings.

The $5\frac{1}{2}$-in.-thick slab is adequate for heavier live loads and for light vehicular wheel loads. It is also strong enough to provide support for light partitions, so that some of the extra footing construction can be eliminated.

For heavy truck loadings, for storage warehouses, and for other situations involving very heavy loads—especially concentrated ones—thicker pavements should be used, although thickness alone is not sufficient, as mentioned previously.

Reinforcement

Thin slabs are ordinarily reinforced with welded wire mesh. The most commonly used meshes are those with a square pattern of wires—typically 4- or 6-in. spacings—with the same wire size in both directions. This reinforcing is generally considered to provide only for shrinkage and temperature effects and to add little to the flexural strength of the slab. The minimum mesh, commonly used with the 3½-in. slab, is a 6 × 6 10/10, which denotes a mesh with No. 10 wires at 6 in. on center in each direction. For thicker slabs the wire gage should be increased or two layers of mesh should be used.

Small-diameter reinforcing bars are also used for slab reinforcement, especially with thicker slabs. These are generally spaced at greater distances than the mesh wires and must be supported during the pouring operation. Unless the slab is actually designed to span, this reinforcement is still considered to function primarily for shrinkage and temperature stress resistance. However, since cracking in the exposed top surface of the slab is usually the most objectionable, specifications usually require the reinforcing to be kept some minimum distance from the top of the slab.

Subbase

The ideal subbase for floor slabs is a well-graded soil, ranging from fine gravel to coarse sand with a minimum of fine materials. This material can usually be compacted to a reasonable density to provide a good structural support, while retaining good drainage properties to avoid moisture concentrations beneath the slab. Where ground water conditions are not critical, this base is usually simply wetted down before pouring the concrete and the concrete is deposited directly on the subbase. The wetting serves somewhat to consolidate the subbase and to reduce the bleeding out of the water and cement from the bottom of the concrete mass.

To reduce further the bleeding-out effect, or where moisture penetration is more critical, a lining membrane is often used between the slab and the sub-grade or base. This usually consists of a 6-mil plastic sheet or a laminated paper-plastic-fiberglass product that possesses considerably more tear resistance and that may be desirable where the construction activity is expected to increase this likelihood.

Joints

Building floor slabs are usually poured in relatively small units, in terms of the horizontal dimension of the slab. The main reason is to control shrinkage cracking. Thus a full break in the slab, formed as a joint between successive pours, provides for the incremental accumulation of the shrinkage effects. Where larger pours are possible or more desirable, control joints are used. These consist of tooled or sawed joints that penetrate some distance down from the finished top surface.

Surface Treatment

Where the slab surface is to serve as the actual wearing surface, the concrete is usually formed to a highly smooth surface by troweling. This surface may then be treated in a number of ways, such as brooming it to make it less slippery, or applying a hardening compound to further toughen the wearing surface. When a separate material—such as tile or a separate concrete fill—is to be applied as the wearing surface, the surface is usually kept deliberately rough. This may be achieved by simply reducing the degree of finished troweling.

Weather Exposure

Once the building is enclosed, interior floor slabs are not ordinarily exposed to exterior weather conditions. In cold climates, however, freezing and extreme temperature ranges should be considered if slabs are exposed to the weather. This may indicate the need for more temperature reinforcing, less distance between control joints, or the use of materials added to the concrete mix to enhance resistance to freezing.

9.10 FRAMED FLOORS ON GRADE

It is sometimes necessary to provide a concrete floor poured directly on the ground in a situation that precludes the use of a simple paving slab and requires a real structural spanning capability of the floor structure. One of these situations is where a deep foundation is provided for support of walls and columns and the potential settlement of upper ground masses may result in a breaking up and subsidence of the paving. Another situation is where considerable fill must be placed beneath the floor, and it is not feasible to produce a compaction of this amount of fill to assure a steady support for the floor.

Figure 9.12 illustrates two techniques that may be used to provide what amounts to a framed concrete slab and beam system poured directly on the ground. Where spans are modest and beam sizes not excessive it may be possible to provide the system in a single pour by simply trenching for the beam forms, as shown in the upper illustration in Fig. 9.12. When large beams must be

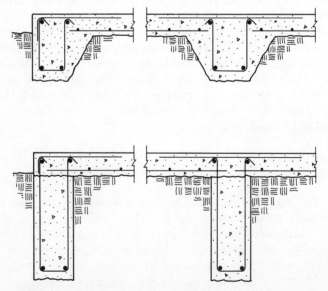

FIGURE 9.12 Details of concrete framed systems cast on grade.

provided, it may be more feasible to use the system shown in the lower part of Fig. 9.12. In this case the stems of the large beams are formed and poured and the slab is poured separately on the fill placed between the beam stems. These two techniques can be blended, of course, with smaller beams trenched in the fill between the large formed beams.

If the system with separately poured beams and slabs is used, it is necessary to provide for the development of shear between the slab and the top of the beam stems. Depending on the actual magnitude of the shear stresses involved, this may be done by various means. If stress is low, it may be sufficient to require a roughening of the surface of the top of the beam stems. If stress is of significant magnitude, shear keys, similar to those used for shear walls, may be used. If stirrups or ties are used in the beam stems, these will extend across the joint and assist in the development of shear.

10

WALLS

Cast concrete is a major means for achieving solid wall construction. In past times masonry was used extensively, and is still competitive in many situations. For structural walls, the two predominant forms are now either cast concrete or masonry with precast concrete units. This chapter presents material relating to structural walls of concrete.

10.1 GENERAL CONCERNS FOR CONCRETE WALLS

Concrete walls are used for a variety of purposes in building construction. Classified with regard to their structural nature, the following types of walls are common.

Bearing Walls, Uniformly Loaded. These may be single story or multistory, carrying loads from floors, roofs, and/or walls above.

Bearing Walls and Concentrated Loads. These are walls that provide support for beams or columns. In most cases they also support uniformly distributed loads.

Basement Walls, Earth Retaining. These are walls that occur at the exterior boundary between interior sublevel spaces and the surrounding earth. In addition to functioning as bearing walls (in most cases) they also span either vertically or horizontally as slabs to resist horizontal earth pressures.

Retaining Walls. This term is usually used to refer to walls that function to achieve grade-level changes, working essentially as vertical cantilevers to resist the horizontal earth pressures from the high side.

Shear Walls. These are walls that are used to brace the building against horizontal (lateral) forces due to wind or earthquakes. The shear referred to is generated in the plane of the wall, as opposed to shear generated in slab-spanning action.

Freestanding Walls. These are walls used as fences or partitions, being supported only at their bases.

Grade Walls. These are walls that occur in buildings without basements; they function to support walls above grade and grade-level floor slabs. They may also function as grade beams or ties in buildings with isolated foundations consisting of column footings, piles, or piers.

It is possible, of course, for walls to serve more than one of these functions. Concrete walls are quite expensive when compared to other types of wall construction and, when used, are usually exploited for all their potential value for structural purposes.

In light building construction, concrete walls are sometimes built without reinforcing. The material in this chapter deals only with reinforced concrete walls.

Regardless of their structural functions, a number of basic considerations apply to all walls. Some considerations of major concern are:

Wall Thickness. Nonstructural walls may be as thin as 4 in.; structural walls must be at least 6 in. thick. In general, slenderness ratio (unsupported height divided by thickness) should not exceed 25. A practical limit for a single pour (total height achieved in one continuous casting) is 15 times the wall thickness; taller walls will require multiple pours. Walls 10 in. or more in thickness should have two layers of reinforcing, one near each wall surface. Basement walls, foundation walls, and party walls must be at least 8 in. thick. Of course, the thickness must also be appropriate to the structural tasks.

Reinforcement. A minimum area of reinforcement equal to 0.0025 times the wall cross section must be provided in a horizontal direction; 0.0015 in a vertical direction. A reduction is possible if bars No. 5 or smaller of grade 60 or higher steel are used. As noted previously, two layers are required for walls 10 in. or more in thickness. The distribution of the total area required between the two layers depends on the wall functions.

Special Reinforcement Requirements. General practice is to provide extra reinforcing at the top, bottom, ends, corners, intersections, and around openings in the wall. Suggested details for placement of reinforcement at these locations is given in the *ACI Detailing Manual* (Ref. 8) and various requirements are given in the ACI Code.

10.2 BEARING WALLS

When the full wall cross section is utilized, bearing strength is limited as follows.
By working stress:

$$P = 0.30f'_c A_1$$

By strength methods:

$$P_u = 0.7(0.85f'_c A_1)$$

When the area developed in bearing is less than the total wall cross section, these loads may be increased by a factor equal to

$\sqrt{A_2/A_1}$, but not more than 2. In this case A_1 is the actual bearing area and A_2 is the area of the full wall cross section.

When the resultant vertical compression force on a wall falls within the middle third of the wall thickness, the wall may be designed as an axial loaded column, using the following empirical formula with the strength method.

$$\phi P_{nw} = 0.55\phi f'_c A_g\left[1 - \left(\frac{l_c}{40h}\right)^2\right]$$

where $\phi = 0.70$

P_{nw} = nominal axial load strength of wall

A_g = the effective area of the wall cross section

l_c = vertical distance between lateral supports

h = overall thickness of the wall

If the wall carries concentrated loads, the length of the wall to be considered as effective for each load shall not exceed the center-to-center distance between the loads nor the actual width of bearing plus four times the wall thickness.

The following example illustrates the procedure for design of a wall with concentrated loads. Design for a uniformly distributed load is essentially the same except that bearing stress and reduced effective area considerations need not be made.

Example. A reinforced concrete bearing wall supports a roof system consisting of precast single tees spaced 8 ft 0 in. on centers. The stem of each T-section is 8 in. wide, but the bearing width is taken as 7 in. to allow for beveled bottom edges. The tees will bear on the full thickness of the wall. The height of the wall is 11 ft 6 in., and the reaction of each tee due to service loads is 22 kips for dead load and 12 kips for live load. Design the wall in accordance with the following specification data: $f'_c = 4000$ psi and $f_y = 40,000$ psi (see Fig. 10.1).

Solution: (1) The factored value of the reaction of one single T-section is

$$P_u = 1.4P_d + 1.7P_l$$

$$P_u = (1.4 \times 22) + (1.7 \times 12) = 51.2 \text{ kips}$$

and

$$\frac{P_u}{\phi} = \frac{51.2}{0.70} = 73.1 \text{ kips}$$

(2) Assume the minimum wall thickness of $h = 6$ in. and check the bearing stress f_b. Letting b' equal the bearing width of the tee stem,

$$f_b = 73{,}100 \div (7 \times 6) = 1740 \text{ psi}$$

$$\text{allowable } f_b = 0.85\phi f_c' = 0.85 \times 0.70 \times 4000 = 2380 \text{ psi}$$

Because f_b is less than the allowable value, bearing on the wall is not critical.

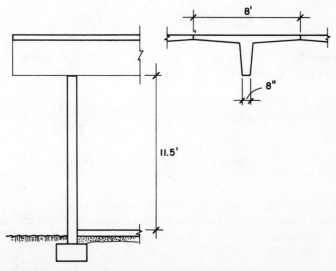

FIGURE 10.1 Construction details for the bearing wall example.

(3) Determine the effective horizontal length of wall. This will be controlled by the bearing width of the T-section plus four times the wall thickness or

$$b' + 4h = 7 + (4 \times 6) = 31 \text{ in.}$$

(4) Check the l_c/h ratio to see that it does not exceed 25.

$$\frac{l_c}{h} = \frac{11.5 \times 12}{6} = 23 < 25 \qquad \text{OK}$$

Therefore, the minimum thickness of $h = 6$ in. is tentatively adopted.

(5) With the full bearing of the T-sections on the wall (and assuming that their deflection will not be sufficient to move the center of reactions outside the middle third of the wall) the "reasonably concentric" loading condition may be considered satisfied.

(6) Determine the allowable capacity of the wall from Code Eq. (14-1). This equation may be written

$$\frac{P_u}{\phi} = 0.55 f'_c A_g \left[1 - \left(\frac{l_c}{40h} \right)^2 \right]$$

Expressing f'_c in kips per square inch, l_c and h in inches, and noting that $A_g = 6 \times 31 = 186$ in.^{2},

$$\frac{P_u}{\phi} = 0.55 \times 4 \times 186 \left[1 - \left(\frac{11.5 \times 12}{40 \times 6} \right)^2 \right]$$

$$\frac{P_u}{\phi} = 409 \times [1 - (0.575)^2] = 274 \text{ kips}$$

which is greater than the required capacity of $P_u/\phi = 73.1$ k determined in step 1. Therefore the capacity of the 6-in.-thick wall is adequate and provides a suitable margin for possible effect of eccentricity.

(7) Select the reinforcement. Because the required amounts of both vertical and horizontal steel are expressed as $\rho \times A_g$, we may work with one linear foot of wall instead of using the effective horizontal length of 31 in.; for this purpose, A_g then becomes $6 \times 12 = 72$ in.^{2}:

$$\text{vertical } A_s = 0.0015 \times 72 = 0.108 \text{ in.}^2 \text{ per linear foot}$$

$$\text{horizontal } A_s = 0.0025 \times 72 = 0.180 \text{ in.}^2 \text{ per linear foot}$$

The maximum spacing of reinforcement in walls is controlled by Section 7.6.5 of the ACI Code which provides that bars shall be spaced not farther apart than three times the wall thickness or more than 18 in. For this wall, the maximum spacing is $3 \times h = 3 \times 6 = 18$ in. Then from Table 6.4

$$\text{vertical steel} = \text{No. 4 bars 18 in. on centers } (A_s = 0.13)$$

$$\text{horizontal steel} = \text{No. 4 bars 12 in. on centers } (A_s = 0.20)$$

both values being expressed as square inches per linear foot of wall.

PROBLEM 10.2.A. An 8-in.-thick reinforced concrete bearing wall is 15 ft high. It supports precast concrete girders 10 ft on centers and each has a service load reaction of 42 k. Of this amount 28 k is due to dead load and 14 k to live load. The girders have full bearing on the wall and the effective width of bearing (parallel to the wall face) may be taken as $7\frac{1}{2}$ in. Determine whether the wall is adequate for this loading and design the required reinforcement. Specification data: $f'_c = 3000$ psi and $f_y = 40,000$ psi.

10.3 BASEMENT WALLS

The identifying characteristic of a basement wall is that most of its height is below grade and it separates building space from earth in contact with its outside face. Such walls must be properly

waterproofed and reinforced to provide for temperature variations as well as to resist bending stresses due to the thrust of the
earth.

Basement walls may or may not be bearing walls, depending
on the structural scheme employed in a particular building. With
respect to the earth-retaining function, a basement wall may be
considered as a slab spanning from column to column or as a slab
with vertical tension reinforcement, the first and basement floor
slabs serving as the two reactions for the horizontal earth pressure. Because the basement height is generally less than the
column spacing, the latter condition occurs most frequently.

Referring to Fig. 10.2a, the earth pressure is considered to be a
horizontal triangular loading with a maximum value at the basement floor and decreasing in magnitude toward the top of the

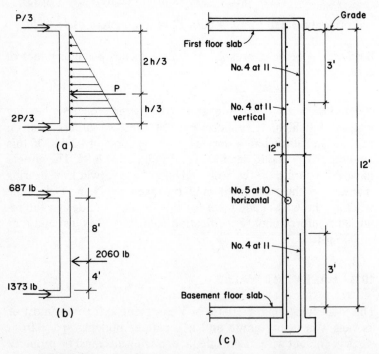

FIGURE 10.2 Considerations for the basement wall example.

slab. The resultant of the earth pressure is represented by P and, when the surface of the retained earth at the top of the wall is horizontal, its magnitude may be determined by the formula

$$P = 0.286 \frac{wh^2}{2}$$

in which w = the weight of the retained earth in pounds per cubic foot and h is the height of the retained earth in feet. (This use of the symbol h to denote height of retained earth or vertical distance between supports should not be confused with its meaning in Sec. 10.2, where it was employed to denote thickness of a bearing wall.) The direction of the resultant earth pressure is horizontal and acts at $\frac{1}{3}h$ from the bottom of the wall slab. The two reactions, or resisting horizontal forces, are $\frac{1}{3}P$ and $\frac{2}{3}P$, the forces resisted by the first floor and basement floor slabs, respectively. For this type of triangular loading the section of the wall slab at which the bending moment is maximum is $0.58 \times h$ from the top of the slab. A wall of this type is in reality a vertical slab with vertical reinforcement. If the wall and the floor slabs are placed at the same time, there is restraint at the two reactions. It is customary in computing the bending moment, however, to consider the slab as simply supported, thus erring on the side of safety. The maximum bending moment for this type of triangular loading is given by the formula

$$M = 0.128WL \qquad \text{or} \qquad M = 0.128Wh$$

where W is the total triangular load and L or h is the span length—the height or vertical distance between supports.

In many instances the required minimum thickness of 8 in. for basement walls, together with the minimum reinforcement ratio of $\rho = 0.0015$ for vertical steel, will satisfy the requirements for bending due to earth pressure. This is frequently the case in residential buildings.

One design procedure applicable to basement walls not supporting significant vertical superimposed loads is to assume a thickness, supply reinforcement in accordance with the minimum ratio, and then investigate this tentative design for adequacy in bending under the lateral load exerted by earth pressure. This procedure is followed in the example presented below. Local knowledge is required to establish a value for w, the weight of the retained earth. This can vary from an average value of approximately 100 lb/ft³ for dry sand to 120 lb/ft³ for wet sand or ordinary wet earth. The local building code should be consulted to ascertain whether mandatory values are specified.

Example. Design a reinforced concrete basement wall 12 ft 0 in. high between the basement and first-floor slabs that serve as supports to take the lateral earth pressure. Exterior finished grade is approximately level with the bottom of the first-floor slab, as indicated in Fig. 10.2c. No appreciable vertical load is carried by the wall because the span of the first-floor construction is parallel to the wall shown in the figure. Specification data: $f'_c = 3000$ psi, $f_y = 40,000$ psi, and the weight of the retained earth is to be taken as 100 lb/ft³,

Solution: (1) Using a 1-ft-wide strip as in slab design, the value of the total earth pressure is

$$P = 0.286 \frac{wh^2}{2} = 0.286 \times \frac{100 \times 12 \times 12}{2} = 2060 \text{ lb}$$

This resultant force acts at $\frac{1}{3}h$, or 4 ft 0 in., from the basement floor slab.

The force resisted by the first floor slab is $\frac{1}{3} \times 2060 = 687$ lb, and the basement floor slab resists a force of $\frac{2}{3} \times 2060 = 1373$ lb (see Fig. 10.2b).

(2) The section of maximum bending moment in the wall slab is located $0.58 \times h$ from the top of the wall, or $0.58 \times 12 = 6.96$ ft, and the magnitude of the moment is

$$M = 0.128Wh = 0.128 \times 2060 \times 12 = 3170 \text{ ft-lb}$$

This represents the service live load moment, and because there is no service dead load moment

$$M_u = 1.7M_l = 1.7 \times 3170 = 5390 \text{ ft-lb}$$

and the required theoretical moment strength is

$$M_t = \frac{M_u}{\phi} = \frac{5390}{0.90} = 5988 \text{ ft-lb}$$

(3) Assume a thickness of 12 in. This gives a gross cross-sectional area A_g of 144 in.^{2} for a 12-in. strip of wall 12 in. thick. If we take minimum $\rho = 0.0015$ as a trial reinforcement ratio,

$$A_s = \rho A_g = 0.0015 \times 144 = 0.216$$

which can be supplied by No. 4 bars spaced 11 in. on centers. This reinforcement is tentatively adopted.

(4) Following the procedure for investigation of rectangular sections (Sec. 6.5), the ultimate value of T (Fig. 6.7) is

$$T = A_s f_y = 0.216 \times 40,000 = 8640 \text{ lb}$$

Setting this value equal to the expression for C (Sec. 6.7) and solving for a, the depth of the rectangular stress block,

$$C = 0.85 f'_c ba = 8640 \text{ lb}$$

and

$$a = \frac{8640}{0.85 f'_c b} = \frac{8640}{0.85 \times 3000 \times 12} = 0.28 \text{ in.}$$

(5) Compute the theoretical moment strength of the 12-in. strip of wall by substituting in formula (2) Sec. 6.5, taking d as $12 - 1$ (for $\frac{3}{4}$ in. concrete cover at inside face) = 11 in.

$$M_t = T\left(d - \frac{a}{2}\right) = 8640\left(11 - \frac{0.28}{2}\right)$$
$$= 93,830 \text{ in.-lb or } 7819 \text{ ft-lb.}$$

Because this value is greater than the required theoretical moment strength of 5988 ft-lb determined in step 2, No. 4 bars on 11-

in. centers are adopted for the vertical reinforcement. These bars are placed near the inner face of the wall.

(6) In addition to vertical bars, horizontal reinforcement must be used. The ACI Code requires that the horizontal reinforcement be a minimum of 0.0025 times the area of the reinforced section of the wall. Therefore $0.0025 \times 12 \times 12 = 0.36$ in.^{2}, and we use No. 5 bars spaced 10 in. on centers.

To provide for possible tensile stresses in the outer portion of the wall at the basement and first floor slabs, No. 4 bars are extended to the fourth points of height, as shown in Fig. 10.2c.

It will be found that neither shear nor development length of reinforcement are critical in this wall. This is commonly the situation for basement walls of this general type and proportions.

For walls over 10 in. in thickness, other than basement walls, Sec. 14.3.4 of the ACI Code requires that both vertical and horizontal reinforcement be placed in two layers. Even though the code exempts basement walls from this requirement, many designers prefer to place two layers in all walls of 12 in. or greater thickness.

Problem 10.3.A. A basement wall 15 ft 0 in. high is supported by the basement and first-floor slabs against the lateral pressure of the exterior earth fill. Design the wall in accordance with the following specification data: $f'_c = 3000$ psi, $f_y = 40,000$ psi, and the weight of the earth fill is to be taken as 100 lb/ft^3.

10.4 RETAINING WALLS

Strictly speaking, any wall that sustains significant lateral soil pressure is a retaining wall. However, the term is usually used with reference to a so-called cantilever retaining wall, which is a freestanding wall without lateral support at its top. For such a wall the major design consideration is for the actual dimension of the ground-level difference that the wall serves to facilitate. The range of this dimension establishes the following different categories for the retaining structure:

Curbs. Curbs are the shortest freestanding retaining structures. The two most common forms are as shown in Fig. 10.3*a*,

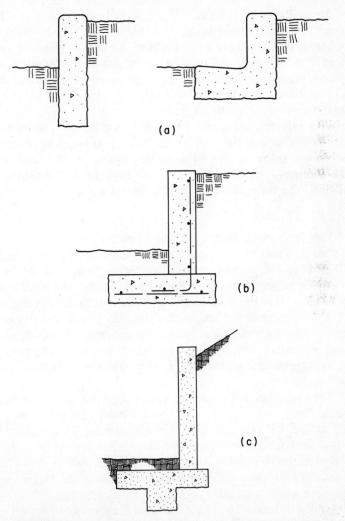

FIGURE 10.3 Typical forms for short retaining structures.

the selection being made on the basis of whether or not it is necessary to have a gutter on the low side of the curb. Use of these structures is typically limited to grade level changes of about 2 ft or less.

Short Retaining Walls.   Vertical walls up to about 10 ft in height are usually built as shown in Fig. 10.3b. These consist of a concrete or masonry wall of uniform thickness. The wall thickness, footing width and thickness, vertical wall reinforcing, and transverse footing reinforcing are all designed for the lateral shear and cantilever bending moment plus the vertical weights of the wall, footing, and earth fill.

When the bottom of the footing is a short distance below grade on the low side of the wall and/or the lateral pressure resistance of the soil is low, it may be necessary to use an extension below the footing—called a shear key—to increase the resistance to sliding. The form of such a key is shown in Fig. 10.3c.

Tall Retaining Walls.   As the wall height increases, it becomes less feasible to use the simple construction shown in Fig. 10.3. The overturning moment increases sharply with the increase in height of the wall. For very tall walls, one modification used is to taper the wall thickness. This permits the development of a reasonable cross section for the high bending stress at the base without an excessive amount of concrete. However, as the wall becomes really tall, it is often necessary to consider the use of various bracing techniques, as shown in the other illustrations in Fig. 10.4.

The design of tall retaining walls is beyond the scope of this book. They should be designed with a more rigorous analysis of the active soil pressure than that represented by the simplified equivalent fluid stress method. In addition, the magnitudes of forces in the reinforced concrete elements of such walls indicate the use of strength design methods rather than the less accurate working stress methods.

Under ordinary circumstances it is reasonable to design relatively short retaining walls by the equivalent fluid pressure

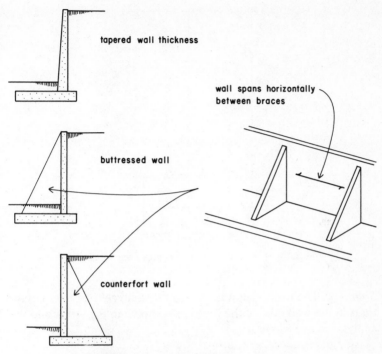

tapered wall thickness

wall spans horizontally
between braces

buttressed wall

counterfort wall

FIGURE 10.4 Forms of tall retaining walls.

method and to use the working stress method for the design of the elements of the wall. The following example illustrates this simplified method of design.

Example. A short retaining wall is proposed with the profile shown in Fig. 10.5. Investigate for the adequacy of the wall dimensions and select reinforcing for the wall and its footing. Use the following data.

Active soil pressure: 30 lb/ft^2 of height.
Soil weight: 100 lb/ft^3.
Maximum allowable soil pressure: 1500 psf.
Concrete strength: $f'_c = 3000$ psi.
Reinforcing: grade 40 bars, $f_y = 40,000$ psi.

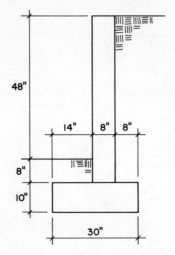

FIGURE 10.5 Form of the retaining wall example.

Solution: The loading condition used to analyze the stress conditions in the wall (above the footing) is shown in Fig. 10.6, and the analysis is as follows.

Maximum lateral pressure:

$$p = 30(4.667 \text{ ft}) = 140 \text{ lb/ft}^2$$

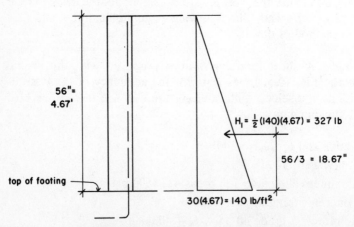

FIGURE 10.6 Lateral load for the wall design.

Total horizontal force:

$$H_1 = \frac{140(4.667)}{2} = 327 \text{ lb}$$

Moment at base of wall:

$$M = 327 \left(\frac{56}{3}\right) = 6104 \text{ in.-lb}$$

For the wall, we assume an approximate effective d of 5.5 in. The tension reinforcing required for the wall is thus

$$A_s = \frac{M}{f_s j d} = \frac{6104}{20,000(0.9)(5.5)} = 0.061 \text{ in.}^2/\text{ft}$$

This may be provided by using No. 3 bars at 20-in. centers, which gives an actual A_s of 0.066 in.^{2}/ft. Since the embedment length of these bars in the footing is quite short, they should be selected conservatively and should have hooks at their ends for additional anchorage.

The loading condition used to investigate the soil stresses and the stress conditions in the footing is shown in Fig. 10.7. In addi-

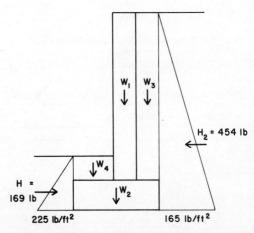

FIGURE 10.7 Loading for the investigation of the footing and the soil stress.

tion to the limit of the maximum allowable soil bearing pressure, it is usually required that the resultant vertical force be kept within the kern limit of the footing. The location of the resultant force is therefore usually determined by a moment summation about the centroid of the footing plan area, and the location is found as an eccentricity from this centroid.

Table 10.1 contains the data and calculations for determination of the location of the resultant force that acts at the bottom of the footing. The position of this resultant is found by dividing the net moment by the sum of the vertical forces, as follows:

$$e = \frac{5793}{1167} = 4.96 \text{ in.}$$

For the rectangular footing plan area, the kern limit will be one-sixth of the footing width or 5 in. The resultant is thus within the kern, and the combined soil stress may be determined by the stress formula as follows:

$$p = \frac{N}{A} \pm \frac{M}{S}$$

where N = total vertical force

A = plan area of the footing

M = net moment about the footing centroid

TABLE 10.1 Determination of the Eccentricity of the Resultant Force

	Force (lb)	Moment arm (in.)	Moment (lb-in.)
H_2	454	22	+9988
w_1	466	3	−1398
w_2	312	0	
w_3	311	11	−3421
w_4	78	8	+624
	$\Sigma_w = 1167$ lb		Net moment = +5793 lb-in.

S = section modulus of the rectangular footing plan
area, which is determined as follows:

$$S = \frac{bh^2}{6} = \frac{1(2.5)^2}{6} = 1.042 \text{ ft}^3$$

The limiting maximum and minimum soil pressures are thus
determined as follows:

$$p = \frac{N}{A} \pm \frac{M}{S} = \frac{1167}{2.5} \pm \frac{5793/12}{1.042} = 467 \pm 463$$
$$= 930 \text{ lb/ft}^2 \text{ maximum and } 4 \text{ lb/ft}^2 \text{ minimum}$$

Since the maximum stress is less than the established limit of
1500 lb/ft^2, vertical soil pressure is not critical for the wall. For
the horizontal force analysis, the procedure varies with different
building codes. The criteria given in this example for soil friction
and passive resistance are those in the *Uniform Building Code*
(Ref. 4) for ordinary sandy soils. This code permits the addition of
these two resistances without modification. Using this data and
technique, the analysis is as follows:

Total active force: 454 lb, as shown in Fig. 10.7.
Friction resistance [(friction factor)(total vertical dead load)]:

$$0.25(1167) = 292 \text{ lb}$$

Passive resistance: 169 lb, as shown in Fig. 10.7.
Total potential resistance:

$$292 + 169 = 461 \text{ lb}$$

Since the total potential resistance is greater than the active
force, the wall is not critical in horizontal sliding.

As with most wall footings, it is usually desirable to select the
footing thickness to minimize the need for tension reinforcing due

to bending. Thus shear and bending stresses are seldom critical, and the only footing stress concern is for the tension reinforcing. The critical section for bending is at the face of the wall, and the loading condition is as shown in Fig. 10.8. The trapezoidal stress distribution produces the resultant force of 833 lb, which acts at the centroid of the trapezoid, as shown in the illustration. Assuming an approximate depth of 6.5 in. for the section, the analysis is as follows:

Moment:

$$M = 833(7.706) = 6419 \text{ in.-lb}$$

Required area:

$$A_s = \frac{M}{f_s j d} = \frac{6149}{20{,}000(0.9)(6.5)} = 0.055 \text{ in.}^2/\text{ft}$$

This requirement may be satisfied by using No. 3 bars at 24-in. centers. For ease of construction, it is usually desirable to have the same spacing for the vertical bars in the wall and the transverse bars in the footing. Thus, in this example, the No. 3 bars at 20-in. centers previously selected for the wall would probably

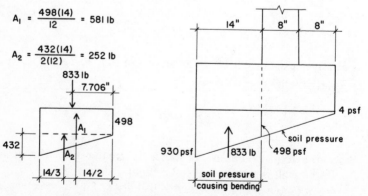

FIGURE 10.8 Bending investigation for the footing.

also be used for the footing bars. The vertical bars can then be held in position by wiring the hooked ends to the transverse footing bars.

Although bond stress is also a potential concern for the footing bars, it is not likely to be critical as long as the bar size is relatively small (less than a No. 6 bar or so).

Reinforcing in the long direction of the footing should be determined in the same manner as for ordinary wall footings. As discussed in Sec. 10.1, we recommend a minimum of 0.15% of the cross section. For the 10-in.-thick and 30-in.-wide footing, this requires

$$A_s = 0.0015(300) = 0.45 \text{ in.}^2$$

We would therefore use three No. 4 bars with a total area of $3(0.2) = 0.6 \text{ in.}^2$.

In most cases designers consider the stability of a short cantilever wall to be adequate if the potential horizontal resistance exceeds the active soil pressure and the resultant of the vertical forces is within the kern of the footing. However, the stability of the wall is also potentially questionable with regard to the usual overturn effect. If this investigation is considered to be necessary, the procedure is as follows.

The loading condition is the same as that used for the soil stress analysis and shown in Fig. 10.7. As with the vertical soil

TABLE 10.2 Analysis for Overturning Effect

	Force (lb)	Moment arm (in.)	Moment (lb-in.)
Overturn:			
H_2	454	22	9988
Restoring moment:			
w_1	466	18	8388
w_2	312	15	4680
w_3	311	26	8086
w_4	78	7	546
			Total: 20,686 lb-in.

stress analysis, the force due to passive soil resistance is not used in the moment calculation, since it is only a potential force. For the overturn investigation, the moments are taken with respect to the toe of the footing. The calculation of the overturning and dead load restoring moments are shown in Table 10.2. The safety factor against overturn is determined as

$$SF = \frac{\text{restoring moment}}{\text{overturning moment}} = \frac{20{,}686}{9988} = 2.07$$

The overturning effect is usually not considered to be critical as long as the safety factor is at least 1.5.

PROBLEM 10.4.A. Design a short retaining wall similar in form to that shown in Fig. 10.5. Referring to the figure, height above grade = 4.5 ft. Use $f'_c = 3$ ksi and grade 40 bars with $f_y = 40$ ksi and $f_s = 20$ ksi.

10.5 SHEAR WALLS

Because of their stiffness, dead weight, and high potential strength, concrete walls are frequently used to brace buildings against the lateral force effects of wind storms and earthquakes. Even when other bracing systems are used, the presence of the very stiff concrete walls makes them attract loading because of their relatively high resistance to deformation.

Design of shear walls for resistance to either wind or seismic effects involves many considerations for building planning, load determination, general lateral resistive system development, anchorage of system components, and investigation of the walls for design. Some of the problems of shear wall planning and design are discussed in Chapter 12, but a full treatment of the topic is beyond the scope of this book. The reader is referred to *Simplified Building Design for Wind and Earthquake Forces* (Ref. 14) for more extensive discussion.

10.6 PRECAST WALLS

Precast concrete walls are most commonly produced by the method of casting them in a flat position at the site. When the concrete is sufficiently strong, they are lifted and placed in the desired position. This method is commonly described as *tilt-up construction,* referring to its early development, when the walls were cast immediately next to their desired location and then literally tilted up into position. Now they are generally lifted and placed with cranes, which allows them to be cast away from the actual desired locations.

Wall units may also be factory-cast, but the practice is more often used for so-called *architectural concrete,* meaning basically nonstructural units, such as those used for curtain wall construction. A critical concern for precast construction is the problem of transporting elements to the building site. Within some reasonable range of distance from a large production facility, it may be possible to obtain many large elements—possibly constituting entire building structures.

Precast walls are typically of conventional reinforced construction. Their structural design in general is subject to the same procedures as those used for cast-in-place construction. A special concern is the provisions that are required to permit lifting and handling of the units. Whether casting is done in a factory or at the building site, it is usually done with the wall unit in a flat position. Picking it up from this position requires development of stress conditions that are not present once the wall is in a vertical position.

10.7 MASONRY WALLS WITH CONCRETE UNITS

As discussed in Sec. 3.11, much of the structural masonry used for building construction today is produced with masonry units of precast concrete (concrete blocks, now called CMUs for "concrete masonry unit"). The two general forms used are described as reinforced or unreinforced. Actually, some reinforcement is typically used in the masonry classified as unreinforced, but it is limited to a type of wire reinforcement laid in the horizontal mortar joints and some steel rods and grout in the voids at critical locations, such as at the ends and tops of walls.

Reinforced masonry with CMUs ordinarily takes the form shown in Fig. 3.4, using units with fewer, larger voids to better accommodate the placing of steel-reinforcement and concrete fill in voids at regular intervals. Code requirements result in these internal reinforced concrete strips at least every 4 ft, both horizontally and vertically, as well as at wall tops, ends, corners, intersections, and on all sides of any wall openings. The result is an extensive reinforced concrete rigid frame built inside the hollow masonry wall.

The general integrity of the masonry, as determined largely by the quality of the mortar and the care and craft of the masons, is of considerably greater concern with unreinforced construction. With reinforced construction, a major component of the structural capacity comes from the reinforced concrete frame inside the wall.

Structural design of unreinforced concrete masonry is done with simple stress investigations and use of allowable stresses, accounting where necessary for the nonsolid cross sections of the hollow construction. Permissible tension stress is quite limited, as with all concrete and masonry in general. Design of reinforced construction generally uses the procedures developed for reinforced concrete, assuming no tensile capacity in the concrete.

Utilization of concrete masonry construction for various situations is illustrated in the building design case studies in Chapter 12. For a full treatment of the various considerations of masonry construction and structural design with masonry, the reader is referred to the references listed at the back of this book. Specific usage of masonry construction varies considerably in different regions of the United States, so that for any design applications the reader is advised to thoroughly investigate local codes and practices.

11

GENERAL CONCERNS FOR BUILDING STRUCTURES

This chapter contains some discussions of general issues relating to design of building structures. These concerns have mostly not been addressed in the presentations in earlier chapters, but require some general consideration when dealing with whole building design situations. General application of these materials is illustrated in the design examples in Chapter 12.

11.1 INTRODUCTION

Materials, methods, and details of building construction vary considerably on a regional basis. There are many factors that affect this situation, including the real effects of response to climate and the availability of construction materials. Even in a single region, differences occur between individual buildings, based on individual styles of architectural design and personal techniques of builders. Nevertheless, at any given time there are usually a few predominant, popular methods of construction that are employed for most buildings of a given type and size. The construction methods and details shown here are reasonable, but in no way are they intended to illustrate a singular, superior style of building.

11.2   DEAD LOADS

Dead load consists of the weight of the materials of which the building is constructed such as walls, partitions, columns, framing, floors, roofs, and ceilings. In the design of a beam, the dead load must include an allowance for the weight of the beam itself. Table 11.1, which lists the weights of many construction materials, may be used in the computation of dead loads. Dead loads are due to gravity and they result in downward vertical forces.

Dead load is generally a permanent load, once the building construction is completed, unless frequent remodeling or rearrangement of the construction occurs. Because of this permanent, long-time, character, the dead load requires certain considerations in design, such as the following:

1. It is always included in design loading combinations, except for investigations of singular effects, such as deflections due to only live load.

2. Its long-time character has some special effects causing sag and requiring reduction of design stresses in wood structures, producing creep effects in concrete structures, and so on.

3. It contributes some unique responses, such as the stabilizing effects that resist uplift and overturn due to wind forces.

TABLE 11.1   Weights of Building Construction

	lb/ft^2	kN/m^2
Roofs		
3-ply ready roofing (roll, composition)	1	0.05
3-ply felt and gravel	5.5	0.26
5-ply felt and gravel	6.5	0.31
Shingles		
Wood	2	0.10

TABLE 11.1 (*Continued*)

	lb/ft^2	kN/m^2
Asphalt	2–3	0.10–0.15
Clay tile	9–12	0.43–0.58
Concrete tile	8–12	0.38–0.58
Slate, $\frac{1}{4}$ in.	10	0.48
Fiberglass	2–3	0.10–0.15
Aluminum	1	0.05
Steel	2	0.10
Insulation		
Fiberglass batts	0.5	0.025
Rigid foam plastic	1.5	0.075
Foamed concrete, mineral aggregate	2.5/in.	0.0047/mm
Wood rafters		
2 × 6 at 24 in.	1.0	0.05
2 × 8 at 24 in.	1.4	0.07
2 × 10 at 24 in.	1.7	0.08
2 × 12 at 24 in.	2.1	0.10
Steel deck, painted		
22 gage	1.6	0.08
20 gage	2.0	0.10
18 gage	2.6	0.13
Skylight		
Glass with steel frame	6–10	0.29–0.48
Plastic with aluminum frame	3–6	0.15–0.29
Plywood or softwood board sheathing	3.0/in.	0.0057/mm
Ceilings		
Suspended steel channels	1	0.05
Lath		
Steel mesh	0.5	0.025
Gypsum board, $\frac{1}{2}$ in.	2	0.10
Fiber tile	1	0.05
Drywall, gypsum board, $\frac{1}{2}$ in.	2.5	0.12
Plaster		
Gypsum, acoustic	5	0.24
Cement	8.5	0.41
Suspended lighting and air distribution systems, average	3	0.15
Floors		
Hardwood, $\frac{1}{2}$ in.	2.5	0.12
Vinyl tile, $\frac{1}{8}$ in.	1.5	0.07
Asphalt mastic	12/in.	0.023/mm

TABLE 11.1 (*Continued*)

	lb/ft^2	kN/m^2
Ceramic tile		
$\frac{3}{4}$ in.	10	0.48
Thin set	5	0.24
Fiberboard underlay, $\frac{3}{8}$ in.	3	0.15
Carpet and pad, average	3	0.15
Timber deck	2.5/in.	0.0047/mm
Steel deck, stone concrete fill, average	35–40	1.68–1.92
Concrete deck, stone aggregate	12.5/in.	0.024/mm
Wood joists		
2 × 8 at 16 in.	2.1	0.10
2 × 10 at 16 in.	2.6	0.13
2 × 12 at 16 in.	3.2	0.16
Lightweight concrete fill	8.0/in.	0.015/mm
Walls		
2 × 4 studs at 16 in., average	2	0.10
Steel studs at 16 in., average	4	0.20
Lath, plaster; *see* Ceilings		
Gypsum dry wall, $\frac{3}{8}$ in. single	2.5	0.12
Stucco, $\frac{7}{8}$ in., on wire and paper or felt	10	0.48
Windows, average, glazing + frame		
Small plane, single glazing, wood or metal frame	5	0.24
Large pane, single glazing, wood or metal frame	8	0.38
Increase for double glazing	2–3	0.10–0.15
Curtain walls, manufactured units	10–15	0.48–0.72
Brick veneer		
4 in., mortar joints	40	1.92
$\frac{1}{2}$ in., mastic	10	0.48
Concrete block		
Lightweight, unreinforced— 4 in.	20	0.96
6 in.	25	1.20
8 in.	30	1.44
Heavy, reinforced, grouted— 6 in.	45	2.15
8 in.	60	2.87
12 in.	85	4.07

11.3 BUILDING CODE REQUIREMENTS

Structural design of buildings is most directly controlled by building codes, which are the general basis for the granting of building permits—the legal permission required for construction. Building codes (and the permit-granting process) are administered by some unit of government: city, county, or state. Most building codes, however, are based on some model code, of which there are three widely used in the United States:

1. The *Uniform Building Code (UBC)* (Ref. 4), which is widely used in the west, as it has the most complete data for seismic design.
2. *The BOCA Basic National Building Code,* used widely in the east and midwest.
3. *The Standard Building Code,* used in the southeast.

These model codes are more similar than different, and are in turn largely derived from the same basic data and standard reference sources, including many industry standards. In the several model codes and many city, county, and state codes, however, there are some items that reflect particular regional concerns.

With respect to control of structures, all codes have materials (all essentially the same) that relate to the following issues:

1. *Minimum Required Live Loads.* This is addressed in Sec. 11.4; all codes have tables similar to those shown in Tables 11.2 and 11.3, which are reproduced from the UBC.
2. *Wind Loads.* These are highly regional in character with respect to concern for local windstorm conditions. Model codes provide data with variability on the basis of geographic zones.
3. *Seismic (Earthquake) Effects.* These are also regional with predominant concerns in the western states. This data, including recommended investigations, is subject to quite frequent modification, as the area of study responds to ongoing research and experience.
4. *Load Duration.* Loads or design stresses are often modified on the basis of the time span of the load, varying from

the life of the structure for dead load to a fraction of a second for a wind gust or a single major seismic shock. Safety factors are frequently adjusted on this basis. Some applications are illustrated in the work in the design examples in this part.

5. *Load Combinations*. These were formerly mostly left to the discretion of designers, but are now quite commonly stipulated in codes, mostly because of the increasing use of ultimate strength design and the use of factored loads.

6. *Design Data for Types of Structures*. These deal with basic materials (wood, steel, concrete, masonry, etc.), specific structures (towers, balconies, pole structures, etc.), and special problems (foundations, retaining walls, stairs, etc.) Industry-wide standards and common practices are generally recognized, but local codes may reflect particular local experience or attitudes. Minimal structural safety is the general basis, and some specified limits may result in questionably adequate performances (bouncy floors, cracked plaster, etc.)

7. *Fire Resistance*. For the structure, there are two basic concerns, both of which produce limits for the construction. The first concern is for structural collapse or significant structural loss. The second concern is for containment of the fire to control its spread. These concerns produce limits on the choice of materials (e.g., combustible or noncombustible) and some details of the construction (cover on reinforcement in concrete, fire insulation for steel beams, etc.)

The work in the design examples in this part is based largely on criteria from the UBC. The choice of this model code reflects only the fact of the degree of familiarity of the author with specific codes in terms of his recent experience.

11.4 LIVE LOADS

Live loads technically include all the nonpermanent loadings that can occur, in addition to the dead loads. However, the term as commonly used usually refers only to the vertical gravity loadings on roof and floor surfaces. These loads occur in combination with the dead loads, but are generally random in character and must be dealt with as potential contributors to various loading combinations, as discussed in Sec. 11.3.

Roof Loads

In addition to the dead loads they support, roofs are designed for a uniformly distributed live load that includes snow accumulation and the general loadings that occur during construction and maintenance of the roof. Snow loads are based on local snowfalls and are specified by local building codes.

Table 11.2 gives the minimum roof live-load requirements specified by the 1988 edition of the UBC. Note the adjustments for roof slope and for the total area of roof surface supported by a structural element. The latter accounts for the increase in probability of the lack of total surface loading as the size of the surface area increases.

TABLE 11.2   Minimum Roof Live Loads

	Minimum Uniformly Distributed Load					
	(lb/ft^2)			(kN/m^2)		
	Tributary Loaded Area for Structural Member					
	(ft^2)			(m^2)		
Roof Slope Conditions	0–200	201–600	Over 600	0–18.6	18.7–55.7	Over 55.7
Flat or rise less than 4 in./ft (1:3); arch or dome with rise less than ⅛ span	20	16	12	0.96	0.77	0.575
Rise 4 in./ft (1:3) to less than 12 in./ft (1:1); arch or dome with rise ⅛ of span to less than ⅜ of span	16	14	12	0.77	0.67	0.575
Rise 12 in./ft (1:1) or greater; arch or dome with rise ⅜ of span or greater	12	12	12	0.575	0.575	0.575
Awnings, except cloth covered	5	5	5	0.24	0.24	0.24
Greenhouses, lath houses, and agricultural buildings	10	10	10	0.48	0.48	0.48

Source: Adapted from the *Uniform Building Code*, 1988 ed. (Ref. 4), copyright © 1988, with the permission of the publishers, the International Conference of Building Officials.

Roof surfaces must also be designed for wind pressure, for which the magnitude and manner of application are specified by local building codes based on local wind histories. For very light roof construction, a critical problem is sometimes that of the upward (suction) effect of the wind, which may exceed the dead load and result in a net upward lifting force.

Although the term *flat roof* is often used, there is generally no such thing; all roofs must be designed for some water drainage. The minimum required pitch is usually $\frac{1}{4}$ in./ft, or a slope of approximately 1:50. With roof surfaces that are close to flat, a potential problem is that of *ponding*, a phenomenon in which the weight of the water on the surface causes deflection of the supporting structure, which in turn allows for more water accumulation (in a pond), causing more deflection, and so on, resulting in an accelerated collapse condition.

Floor Loads

The live load on a floor represents the probable effects created by the occupancy. It includes the weights of human occupants, furniture, equipment, stored materials, and so on. All building codes provide minimum live loads to be used in the design of buildings for various occupancies. Since there is a lack of uniformity among different codes in specifying live loads, the local code should always be used. Table 11.3 contains values for floor live loads as given by the UBC.

Although expressed as uniform loads, code-required values are usually established large enough to account for ordinary concentrations that occur. For offices, parking garages, and some other occupancies, codes often require the consideration of a specified concentrated load as well as the distributed loading. Where buildings are to contain heavy machinery, stored materials, or other contents of unusual weight, these must be provided for individually in the design of the structure.

When structural framing members support large areas, most codes allow some reduction in the total live load to be used for design. These reductions, in the case of roof loads, are incorporated into the data in Table 11.2. The following is the method given in the UBC for determining the reduction permitted for beams, trusses, or columns that support large floor areas.

Except for floors in places of assembly (theaters, etc.), and except for live loads greater than 100 psf [4.79 kN/m^2], the design live load on a member may be reduced in accordance with the formula

$$R = 0.08 (A - 150)$$

$$[R = 0.86 (A - 14)]$$

The reduction shall not exceed 40% for horizontal members or for vertical members receiving load from one level only, 60% for

TABLE 11.3 Minimum Floor Loads

Use or Occupancy		Uniform Load		Concentrated Load	
Description	Description	psf	kN/m^2	lb	kN
Armories		150	7.2		
Assembly areas and auditoriums and balconies therewith	Fixed seating areas	50	2.4		
	Movable seating and other areas	100	4.8		
	Stages and enclosed platforms	125	6.0		
Cornices, marquees, and residential balconies		60	2.9		
Exit facilities		100	4.8		
Garages	General storage, repair	100	4.8	a	
	Private pleasure car	50	2.4	a	
Hospitals	Wards and rooms	40	1.9	1000	4.5
Libraries	Reading rooms	60	2.9	1000	4.5
	Stack rooms	125	6.0	1500	6.7
Manufacturing	Light	75	3.6	2000	9.0
	Heavy	125	6.0	3000	13.3
Offices		50	2.4	2000	9.0
Printing plants	Press rooms	150	7.2	2500	11.1
	Composing rooms	100	4.8	2000	9.0
Residential		40	1.9		
Rest rooms		b			
Reviewing stands, grandstands, and bleachers		100	4.8		
Roof decks (occupied)	Same as area served				
Schools	Classrooms	40	1.9	1000	4.5
Sidewalks and driveways	Public access	250	12.0	a	
Storage	Light	125	6.0		
	Heavy	250	12.0		
Stores	Retail	75	3.6	2000	9.0
	Wholesale	100	4.8	3000	13.3

Source: Adapted from the *Uniform Building Code*, 1988 ed. (Ref. 4), copyright © 1988, with the permission of the publishers, the International Conference of Building Officials.

a Wheel loads related to size of vehicles that have access to the area.

b Same as the area served or minimum of 50 psf.

other vertical members, nor R as determined by the formula

$$R = 23.1 \left(1 + \frac{D}{L}\right)$$

In these formulas

R = reduction, in percent

A = area of floor supported by a member

D = unit dead load/sq ft of supported area

L = unit live load/sq ft of supported area

In office buildings and certain other building types, partitions may not be permanently fixed in location but may be erected or moved from one position to another in accordance with the requirements of the occupants. In order to provide for this flexibility, it is customary to require an allowance of 15 to 20 psf [0.72 to 0.96 kN/m²], which is usually added to other dead loads.

11.5 LATERAL LOADS

As used in building design, the term *lateral load* is usually applied to the effects of wind and earthquakes, as they induce horizontal forces on stationary structures. From experience and research, design criteria and methods in this area are continuously refined, with recommended practices being presented through the various model building codes, such as the UBC.

Space limitations do not permit a complete discussion of the topic of lateral loads and design for their resistance. The following discussion summarizes some of the criteria for design in the latest edition of the UBC. Examples of application of these criteria are given in the chapters that follow containing examples of building structural design. For a more extensive discussion the reader is referred to *Simplified Building Design for Wind and Earthquake Forces* (Ref. 14).

Wind

Where wind is a major local problem, local codes are usually more extensive with regard to design requirements for wind. However, many codes still contain relatively simple criteria for wind design. One of the most up-to-date and complex standards for wind design is contained in the *Minimum Design Loads for Buildings and Other Structures,* ANSI A58.1-1982 (Ref. 10), published by the American National Standards Institute in 1982.

Complete design for wind effects on buildings includes a large number of both architectural and structural concerns. The following is a discussion of some of the requirements for wind as taken from the 1988 edition of the UBC, which is in general conformance with the material presented in the ANSI Standard just mentioned.

Basic Wind Speed. This is the maximum wind speed (or velocity) to be used for specific locations. It is based on recorded wind histories and adjusted for some statistical likelihood of occurrence. For the continental United States the wind speeds are taken from UBC, Fig. No. 4. As a reference point, the speeds are those recorded at the standard measuring position of 10 m (approximately 33 ft) above the ground surface.

Exposure. This refers to the conditions of the terrain surrounding the building site. The ANSI Standard describes four conditions (A, B, C, and D), although the UBC uses only two (B and C). Condition C refers to sites surrounded for a distance of one-half mile or more by flat, open terrain. Condition B has buildings, forests, or ground-surface irregularities 20 ft or more in height covering at least 20% of the area for a distance of 1 mile or more around the site.

Wind Stagnation Pressure (q_s). This is the basic reference equivalent static pressure based on the critical local wind speed. It is given in UBC Table No. 23-F and is based on the following formula as given in the ANSI Standard:

$$q_s = 0.00256V^2$$

Example. For a wind speed of 100 mph

$$q_s = 0.00256V^2 = 0.00256(100)^2$$
$$= 25.6 \text{ psf } [1.23 \text{ kPa}]$$

which is rounded off to 26 psf in the UBC table.

Design Wind Pressure. This is the equivalent static pressure to be applied normal to the exterior surfaces of the building and is determined from the formula (UBC Formula 11-1, Section 2311)

$$p = C_e C_q q_s I$$

where p = design wind pressure in psf

C_e = combined height, exposure, and gust factor coefficient as given in UBC Table No. 23-G

C_q = pressure coefficient for the structure or portion of structure under consideration as given in UBC Table No. 23-H

q_s = wind stagnation pressure at 30 ft given in UBC Table No. 23-F

I = importance factor

The importance factor is 1.15 for facilities considered to be essential for public health and safety (such as hospitals and government buildings) and buildings with 300 or more occupants. For all other buildings the factor is 1.0.

The design wind pressure may be positive (inward) or negative (outward, suction) on any given surface. Both the sign and the value for the pressure are given in the UBC table. Individual building surfaces, or parts thereof, must be designed for these pressures.

Design Methods. Two methods are described in the Code for the application of the design wind pressures in the design of structures. For design of individual elements particular values are given in UBC Table 23-H for the C_q coefficient to be used in determining p. For the primary bracing system the C_q values and their use is to be as follows:

Method 1 (Normal Force Method). In this method wind pressures are assumed to act simultaneously normal to all exterior surfaces. This method is required to be used for gabled rigid frames and may be used for any structure.

Method 2 (Projected Area Method). In this method the total wind effect on the building is considered to be a combination of a single inward (positive) horizontal pressure acting on a vertical surface consisting of the projected building profile and an outward (negative, upward) pressure acting on the full projected area of the building in plan. This method may be used for any structure less than 200 ft in height, except for gabled rigid frames. This is the method generally employed by building codes in the past.

Uplift. Uplift may occur as a general effect, involving the entire roof or even the whole building. It may also occur as a local phenomenon such as that generated by the overturning moment on a single shear wall. In general, use of either design method will account for uplift concerns.

Overturning Moment. Most codes require that the ratio of the dead load resisting moment (called the restoring moment, stabilizing moment, etc.) to the overturning moment be 1.5 or greater. When this is not the case, uplift effects must be resisted by anchorage capable of developing the excess overturning moment. Overturning may be a critical problem for the whole building, as in the case of relatively tall and slender tower structures. For buildings braced by individual shear walls, trussed bents, and rigid-frame bents, overturning is investigated for the individual bracing units. Method 2 is usually used for this investigation, except for very tall buildings and gabled rigid frames.

Drift. Drift refers to the horizontal deflection of the structure due to lateral loads. Code criteria for drift are usually limited to requirements for the drift of a single story (horizontal movement of one level with respect to the next above or below). The UBC does not provide limits for wind drift. Other standards give various recommendations, a common one being a limit of story drift to 0.005 times the story height (which is the UBC limit for seismic drift). For masonry structures wind drift is sometimes limited to 0.0025 times the story height. As in other situations involving structural deformations, effects on the building construction must be considered; thus the detailing of curtain walls or interior partitions may affect limits on drift.

Combined Loads.   Although wind effects are investigated as isolated phenomena, the actions of the structure must be considered simultaneously with other phenomena. The requirements for load combinations are given by most codes, although common sense will indicate the critical combinations in most cases. With the increasing use of load factors the combinations are further modified by applying different factors for the various types of loading, thus permitting individual control based on the reliability of data and investigation procedures and the relative significance to safety of the different load sources and effects. Required load combinations are described in Sec. 2303 of the UBC.

Special Problems.   The general design criteria given in most codes are applicable to ordinary buildings. More thorough investigation is recommended (and sometimes required) for special circumstances such as the following:

Tall Buildings. These are critical with regard to their height dimension as well as the overall size and number of occupants inferred. Local wind speeds and unusual wind phenomena at upper elevations must be considered.

Flexible Structures. These may be affected in a variety of ways, including vibration or flutter as well as the simple magnitude of movements.

Unusual Shapes. Open structures, structures with large overhangs or other projections, and any building with a complex shape should be carefully studied for the special wind effects that may occur. Wind-tunnel testing may be advised or even required by some codes.

Use of code criteria for various ordinary buildings is illustrated in the design examples in Chapter 12.

Earthquakes

During an earthquake a building is shaken up and down and back and forth. The back-and-forth (horizontal) movements are typically more violent and tend to produce major unstabilizing effects on buildings; thus structural design for earthquakes is mostly done in terms of considerations for horizontal (called lateral) forces. The lateral forces are actually generated by the weight of the building—or, more specifically, by the mass of the building that represents both an inertial resistance to movement and the source for kinetic energy once the building is actually in motion. In the simplified procedures of the equivalent static force method, the building structure is considered to be loaded by a set of horizontal forces consisting of some fraction of the building weight. An analogy would be to visualize the building as being rotated vertically 90° to form a cantilever beam, with the ground as the fixed end and with a load consisting of the building weight.

In general, design for the horizontal force effects of earthquakes is quite similar to design for the horizontal force effects of wind. Indeed, the same basic types of lateral bracing (shear walls, trussed bents, rigid frames, etc.) are used to resist both force effects. There are indeed some significant differences, but in the main a system of bracing that is developed for wind bracing will most likely serve reasonably well for earthquake resistance as well.

Because of its considerably more complex criteria and procedures, we have chosen not to illustrate the design for earthquake effects in the examples in this part. Nevertheless, the development of elements and systems for the lateral bracing of the buildings in the design examples here is quite applicable in general to situations where earthquakes are a predominant concern. For structural investigation, the principal difference is in the determination of the loads and their distribution in the building. Another major difference is in the true dynamic effects, critical wind force being usually represented by a single, major, one-direction punch from a gust, while earthquakes represent rapid back-and-forth, reversing-direction actions. However, once the dynamic effects are translated into equivalent static forces, design concerns for the bracing systems are very similar, involving considerations for shear, overturning, horizontal sliding, and so on.

For a detailed explanation of earthquake effects and illustrations of the investigation by the equivalent static force method the reader is referred to *Simplified Building Design for Wind and Earthquake Forces* (Ref. 14).

11.6 STRUCTURAL PLANNING

Planning a structure requires the ability to perform two major tasks. The first is the logical arranging of the structure itself, regarding its geometric form, its actual dimensions and proportions, and the ordering of the elements for basic stability and reasonable interaction. All of these issues must be faced, whether the building is simple or complex, small or large, of ordinary construction or totally unique. Spanning beams must be supported and have depths adequate for the spans; thrusts of arches must be resolved; columns above should be centered over columns below; and so on.

The second major task in structural planning is the development of the relationships between the structure and the building in general. The building plan must be "seen" as a structural plan. The two may not be quite the same, but they must fit together. "Seeing" the structural plan (or possibly alternative plans) inherent in a particular architectural plan is a major task for designers of building structures.

Hopefully, architectural planning and structural planning are done interactively, not one after the other. The more the architect knows about the structural problems and the structural designer (if another person) knows about architectural problems, the more likely it is possible that an interactive design development may occur.

Although each individual building offers a unique situation if all of the variables are considered, the majority of building design problems are highly repetitious. The problems usually have many alternative solutions, each with its own set of pluses and minuses in terms of various points of comparison. Choice of the final design involves the comparative evaluation of known alternatives and the eventual selection of one.

The word *selection* may seem to imply that all the possible solutions are known in advance, not allowing for the possibility of a new solution. The more common the problem, the more this may be virtually true. However, the continual advance of science and technology and the fertile imagination of designers make new solutions an ever-present possibility, even for the most common problems. When the problem is truly a new one in terms of a new building use, a scale jump, or a new performance situation, there is a real need for innovation. Usually, however, when new solutions to old problems are presented, their merits must be compared to established previous solutions in order to justify them. In its broadest context the selection process includes the consideration of all possible alternatives: those well known, those new and unproven, and those only imagined.

11.7 SYSTEMS INTEGRATION

Good structural design requires integration of the structure into the whole physical system of the building. It is necessary to realize the potential influences of structural design decisions on the general architectural design and on the development of the systems for power, lighting, thermal control, ventilation, water supply, waste handling, vertical transportation, firefighting, and so on. The most popular structural systems have become so in many cases largely because of their ability to accommodate the other subsystems of the building and to facilitate popular architectural forms and details.

11.8 ECONOMICS

Dealing with dollar cost is a very difficult, but necessary, part of structural design. For the structure itself, the bottom-line cost is the delivered cost of the finished structure, usually measured in units of dollars per square foot of the building. For individual components, such as a single wall, units may be used in other forms. The individual cost factors or components, such as cost of materials, labor, transportation, installation, testing, and inspection, must be aggregated to produce a single unit cost for the entire structure.

Designing for control of the cost of the structure is only one aspect of the design problem, however. The more meaningful cost is that for the entire building construction. It is possible that certain cost-saving efforts applied to the structure may result in increases of cost of other parts of the construction. A common example is that of the floor structure for multistory buildings. Efficiency of floor beams occurs with the generous provision of beam depth in proportion to the span. However, adding inches to beam depths with the unchanging need for dimensions required for floor and ceiling construction and installation of ducts and lighting elements means increasing the floor-to-floor distance and the overall height of the building. The resulting increases in cost for the added building skin, interior walls, elevators, piping, ducts, stairs, and so on, may well offset the small savings in cost of the beams. The really effective cost-reducing structure is often one that produces major savings of nonstructural costs, in some cases at the expense of less structural efficiency.

Real costs can only be determined by those who deliver the completed construction. Estimates of cost are most reliable in the form of actual offers or bids for the construction work. The farther the cost estimator is from the actual requirement to deliver the goods, the more speculative the estimate. Designers, unless they are in the actual employ of the builder, must base any cost estimates on educated guesswork deriving from some comparison with similar work recently done in the same region. This kind of guessing must be adjusted for the most recent developments in terms of the local markets, competitiveness of builders and suppliers, and the general state of the economy. Then the four best guesses are placed in a hat and one is drawn out.

Serious cost estimating requires a lot of training and experience and an ongoing source of reliable, timely information. For major projects various sources are available, in the form of publications or computer hookups.

The following are some general rules for efforts that can be made in the structural design work in order to have an overall, general cost-saving attitude.

1. Reduction of material volume is usually a means of reducing cost. However, unit prices for different grades must be noted. Higher grades of steel or wood may be proportionally more expensive than the higher stress values they represent; more volume of cheaper material may be less expensive.

2. Use of standard, commonly stocked products is usually a cost savings, as special sizes or shapes may be premium priced. Wood 2×3 studs may be higher in price than 2×4 studs, since the 2×4 is so widely used and bought in large quantities.

3. Reduction in the complexity of systems is usually a cost savings. Simplicity in purchasing, handling, managing of inventory, and so on, will be reflected in lower bids as builders anticipate simpler tasks. Use of the fewest number of different grades of materials, sizes of fasteners, and other such variables is as important as the fewest number of different parts. This is especially true for any assemblage done on the building site; large inventories may not be a problem in a factory, but usually are on the site.

4. Cost reduction is usually achieved when materials, products, and construction methods are highly familiar to local builders and construction workers. If real alternatives exist, choice of the "usual" one is the best course.

5. Do not guess at cost factors; use real experience, yours or others. Costs vary locally, by job size, and over time. Keep up to date with cost information.

6. In general, labor cost is greater than material cost. Labor for building forms, installing reinforcement, pouring, and finishing concrete surfaces is *the* major cost factor for site-poured concrete. Savings in these areas are much more significant than saving of material volume.

7. For buildings of an investment nature, time is money. Speed of construction may be a major advantage. However, getting the structure up fast is not a true advantage unless the other aspects of the construction can take advantage of the time gained. Steel frames often go up quickly, only to stand around and rust while the rest of the work catches up.

12

BUILDING STRUCTURES: DESIGN EXAMPLES

This chapter presents examples of the design of structural systems for buildings. The buildings used for the examples have been chosen to create a range of situations in order to be able to demonstrate the use of various structural components. Design of individual elements of the structural systems is largely based on materials presented in the earlier chapters. To conserve space, reference is made to computations demonstrated in the earlier chapters, so that a few actual computations are demonstrated here. What is of primary concern in this chapter is the demonstration of the design process for whole systems, and consideration of the many factors that influence design decisions. Many of these factors are not structural in nature, but nevertheless have major influences on the final form and details of the structure.

12.1 BUILDING ONE

Building One consists of a one-story, box-shaped structure that is intended for commercial occupancy. For maximum flexibility in terms of interior rearrangements, it is desired to have a clear-span roof structure. Figure 12.1 shows a scheme for the building that uses concrete masonry walls and roof trusses of open-web steel joists. We assume the following design data:

Roof live load: 20 psf (reducible).

Design wind pressure: 20 psf (assumed, UBC method 2).

CMUs: medium-weight units, grade N, ASTM C90, $f'_m = 1350$ psi, mortar type S.

The general profile of the building is shown in Fig. 12.1c, which indicates a low slope roof, a flat ceiling, and a short parapet at the exterior walls. The general form of the construction is shown in the wall section in Fig. 12.1d. Lateral bracing is achieved with perimeter shear walls, which take the form of individual, vertically cantilevered piers. Design for lateral loads is discussed in Sec. 12.3. We first consider the design for gravity loads.

12.2 BUILDING ONE: DESIGN FOR GRAVITY LOADS

The masonry walls serve as bearing walls to support the roof construction. For the dead weight of the roof construction we determine the following (see Table 11.1):

Three-ply felt and gravel roofing	= 5.5 psf
Foamed concrete insulating fill, 4 in.	= 10.0
Formed sheet steel deck, 20 gage	= 2.0
Open-web joists, from supplier's catalog	= 12.0
Ceiling: wood nailers and blocking	= 1.0
gypsum drywall	= 2.5
Lighting, HVAC, etc.	= 3.0
Total roof dead load	= 36.0 psf

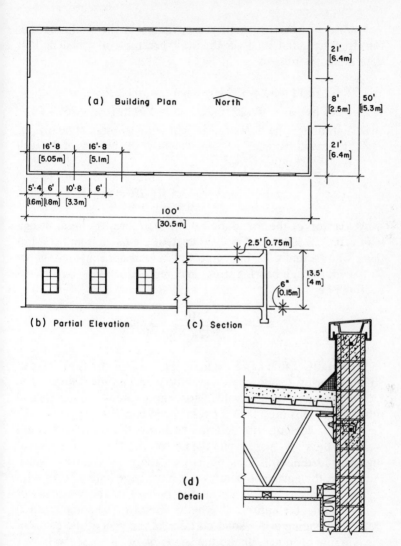

FIGURE 12.1 Building One.

Ignoring for the moment the openings in the wall, the uniformly distributed load on the wall per foot of wall length is determined as follows:

1. Roof dead load = (25 ft)(36 psf) = 900 lb/ft.
2. Roof live load = (25 ft)(20 psf) = 500 lb/ft (not reduced).
3. Estimating the wall dead weight at an average of 60 psf per ft^2 of wall surface, the total weight at the bottom of the wall is

$$\text{wall dead load} = (13.5 \text{ ft})(60) = 810 \text{ lb/ft}$$

At the top of the foundation wall, therefore, the total design load is 900 + 500 + 810 = 2210 lb/ft. If we use a nominal 8-in.-thick wall (actually 7.5 in., typically), and assume a block with a 50% void, the unit bearing stress in compression at the base of the wall will be

$$f_p = \frac{P}{A} = \frac{2210}{7.5(12)(0.50)} = 49 \text{ psi}$$

From UBC Table 24-H we find an allowable stress of 160 psi, so the bearing condition for gravity loads is not critical. The referenced table yields values for unreinforced masonry, and other provisions are made for reinforced construction.

The wall loading just determined is for the north and south walls, as the roof trusses span the narrow dimension of the building. The loading will thus be less on the east and west walls. However, the north and south walls have large openings for windows, so the actual bearing at the bottom of the walls is not uniform along the entire wall length. There will be some concentration of loading in the solid portions of the wall at the edges of the window openings, due to the spanning of the upper wall over the openings. The details of the construction for achieving the 6-ft span over the openings and resisting the load concentration at the edge of the openings will depend partly on the form of the masonry.

If the masonry is reinforced—that is, block voids are vertically aligned and voids are filled with reinforcement and concrete at regular intervals—a lintel over the opening will be developed as a horizontally reinforced beam and the edges of the openings will be developed as reinforced columns. If unreinforced construction is used, it is most common to use a steel lintel over the opening, although a sitecast or precast concrete lintel could also be integrated into the block arrangement.

The section in Fig. 12.2 shows the use of a short foundation wall and a footing for support of the wall. If a concrete masonry wall is used, as shown in Fig. 12.2, it will ordinarily be constructed as shown, with all block voids filled with concrete and some horizontal reinforcement in the top of the wall. This wall will generally serve to distribute the gravity loads in a uniform manner to the footing, with an assumed loading condition as shown in Fig. 12.3. The reinforcement in the top is thus useful for

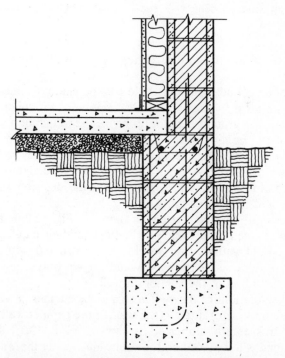

FIGURE 12.2 Building One: masonry foundation wall.

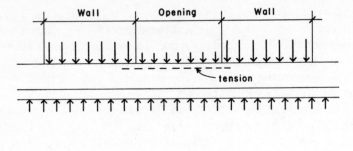

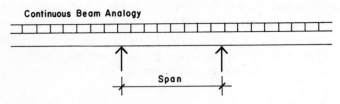

FIGURE 12.3 Building One: spanning action of the foundations.

resistance of the tension occurring as the wall spans between openings.

The short foundation wall could also be made of sitecast concrete, as shown in Fig. 12.4b. Or, as is common when severe freezing of the ground is not a problem, the foundation wall and footing could be combined into a single element, called a grade beam, as shown in Fig. 12.4a. Other factors, such as soil condition, roof drainage systems, building underground utilities, or structures for site plantings, may affect the decision as to the most desired form of construction.

If a separate footing is used, as shown in Fig. 12.2 or 12.4b, it could be selected from Table 9.3. For this building, a quite narrow footing would suffice, and would be constructed without any transverse reinforcement, as it would project only a few inches beyond the edges of the supported wall.

A consideration for the wall design is the manner in which the roof loading is transmitted to the wall. If the parapet wall is used and the trusses are supported as shown in Fig. 12.1f, the roof gravity load is not placed on the centroidal axis of the wall and a bending is induced in the wall due to its eccentricity. The wall

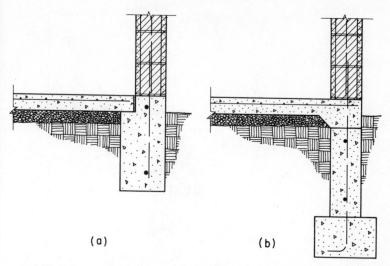

(a)   (b)

FIGURE 12.4 Building One: cast concrete options for the foundations.

must be designed for the combined bending and compression, as discussed in Sec. 10.7. The procedures for this depend on the form of the wall construction—as reinforced or unreinforced.

Additionally, of course, the walls must be designed for lateral effects of wind or earthquakes, as discussed in the next section. In the end, the wall design must satisfy the various possible critical loading combinations, as discussed in Sec. 11.3.

A modification of the construction that generally eliminates the bending caused by gravity loading is shown in Fig. 12.5. Here the roof trusses bear on top of the wall and a short cantilevered roof edge and soffit are developed.

12.3 BUILDING ONE: DESIGN FOR LATERAL LOADS

Design for the effects of wind or earthquakes begins with some basic decisions about the form of the lateral bracing system. For this building consisting only of a perimeter wall structure and a flat roof, the simplest system is that using the roof deck as a horizontal diaphragm in combination with a vertical bracing sys-

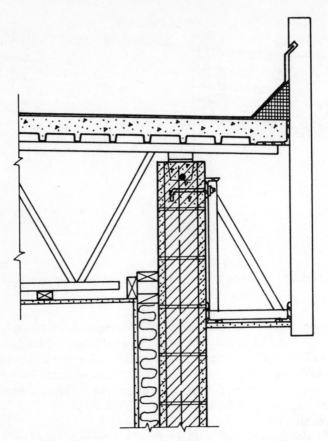

FIGURE 12.5 Building One: alternative for the roof-to-wall detail. For comparison, see Fig. 12.1*d*.

tem consisting of the perimeter walls. If the roof deck is not capable of the required diaphragm actions, or is considerably discontinuous because of openings or geometric variations, it may be necessary to develop some other form of horizontal bracing, such as a horizontal truss system. However, in this case, for the size of the building, the formed sheet steel deck is most likely adequate for the required diaphragm tasks.

The perimeter bracing could consist of units of moment-resistive rigid frames or trussed bents. In this case, however, the structural masonry walls are more than sufficient. Depending on details of the wall construction, the walls may serve as lateral bracing in their own planes by one of two means, as shown in Fig. 12.6. In Fig. 12.6a the walls are shown as consisting of individual piers acting as vertical cantilevers with fixed bases. This action is developed by considering the openings for the windows to constitute complete breaks in the wall's continuous nature. This could be achieved by construction joints in the masonry or by a complete change to another form of construction between the masonry piers.

Another form of action for the walls in resisting lateral forces is shown in Fig. 12.6b. In this case, the wall is essentially considered to act as a continuous rigid frame bent. The deep, stiff, continuous strips of masonry above and below the windows are considered to be nonflexing elements, and the piers of masonry between the windows act as columns, fixed at their tops and bottoms. If the masonry is built as reinforced construction, and the portions above and below the windows are indeed quite deep, this is a reasonable assumption for the wall action.

For the individual pier action shown in Fig. 12.6a, the single wall units will be designed as individual, freestanding shear walls. This involves considerations for the following:

1. *Horizontal Shear in the Wall.* This unit stress is determined simply by dividing the horizontal force by the horizontal cross section of the wall. An allowable stress (or ultimate design strength) is established, and the materials, details, and any necessary reinforcing of the wall is determined.

2. *Bending (Cantilever Beam Action) of the Wall.* Depending on the wall form and construction, the bending may be resisted by the entire wall, or developed essentially by the two opposed wall ends acting like flanges of an I-beam. For reinforced masonry with concrete blocks, the latter is usually assumed, with the two end "columns" acting in compression and tension.

3. *Anchorage of the Wall for Overturning Effect (Rotation at the Base).* The basis for this investigation is shown in Fig.

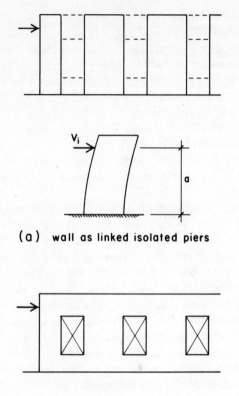

(a) wall as linked isolated piers

(b) continuous wall

FIGURE 12.6 Building One: functioning of the masonry shear walls.

12.7, using the criteria from the 1988 UBC (Ref. 4) for consideration of wind or seismic effects.

4. *Development of the Force Transfers.* This involves the considerations of the construction details to achieve the transfer of force from the roof diaphragm to the piers and from the piers to their supports. Attachment of the roof framing to the wall will achieve the roof-to-wall transfer. Ordinarily, the doweling of the vertical foundation by reinforcement (see Fig. 12.2 or 12.4b) will achieve the necessary base tiedown effect (T in Fig. 12.7).

For any form of construction there are usually some minimum requirements that establish a base level of capability once the form of construction is chosen. For structural masonry, the usual code requirements for the masonry units, mortar, and some details of the construction will establish this minimal construction. In many applications, for buildings of modest size, the structural capacity of this minimal construction will not be exceeded. Thus nothing extra need be done to develop the required structural actions. Such is the case for this building. If properly detailed and built to code standards, either reinforced or unreinforced masonry could achieve the building shown in Fig. 12.1, with no significant "extras" required.

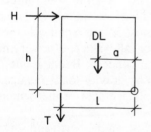

To determine T:

for wind – $\quad DL(a) + T(l) = 1.5\,[H(h)]$

for seismic – $\quad 0.85\,[DL(a)] + T(l) = H(h)$

FIGURE 12.7 Determination of stability and tiedown requirements for a shear wall; working stress method. See UBC Sections 2311(e) and 2312(h)1.

Masonry shear walls are ordinarily quite heavy, so that many have sufficient dead weight to counteract the overturning effects for shear wall action (Fig. 12.7). Thus the tiedown provided by foundation dowels is a bonus, but not really required for the lateral force resistance.

Figure 12.8 shows the basis for consideration of the horizontal force effects of wind on Building One. Uplift on the roof must also be considered, but the effects illustrated in Fig. 12.8 are the usual basis for design of the roof deck as a horizontal diaphragm and the determination of the lateral loads on the perimeter bracing.

The walls must also act as spanning elements in resisting the direct wind pressures on their surfaces. As shown in Fig. 12.8a, the walls span from floor to roof, acting in one of the two ways shown. If the construction is as shown in Fig. 12.1d, and the wall is continuous past the roof edge, it will act as a beam with an overhanging end. If the construction is as shown in Fig. 12.4, the parapet is developed in conjunction with the roof construction, and the wall spans simply between the floor and roof levels.

In either of the cases shown in Fig. 12.8a, some of the wind load on the wall goes directly into the edge of the floor and is not delivered to the edge of the roof diaphragm. Thus the wind load to be used for the investigation of the roof and the shear walls is that shown in Fig. 12.8b (for east or west direction wind), or that shown in Fig. 12.8c (for north or south direction wind).

The spanning action of the walls (Fig. 12.8a) results in bending in the walls, which must be combined with the actions required by the gravity loads. For tall walls and/or very high wind pressures, this may be a critical concern. Bending due to wind may add to any bending caused by gravity loads (as discussed in Sec. 12.2) and require some enhancement of the bending resistance of the construction.

Considering the wind on the north and south walls, and assuming a wall action as shown for case 2 in Fig. 12.8a, the wind load delivered to the roof edge is

$$(20 \text{ psf}) \frac{10.5}{2} + (20 \text{ psf})(2.5) = 155 \text{ lb/ft}$$

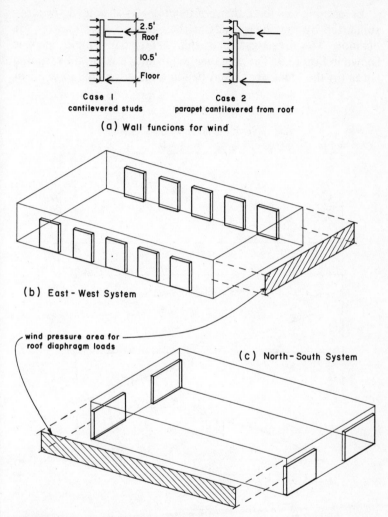

FIGURE 12.8 Building One: considerations of wind for the design of the exterior masonry walls.

In resisting this load, the roof functions as a spanning member supported by the shear walls at the east and west ends of the building. The investigation of this 100-ft span simple beam is shown in Fig. 12.9. The roof diaphragm must sustain the resulting shear (in the steel deck) and bending (as edge compression or

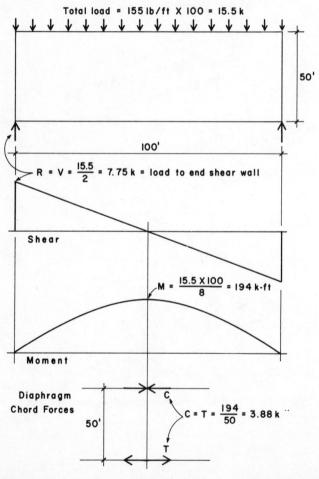

FIGURE 12.9 Building One: functions of the roof structure as a horizontal diaphragm.

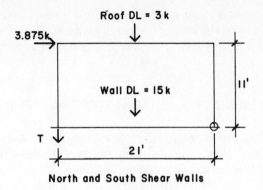

North and South Shear Walls

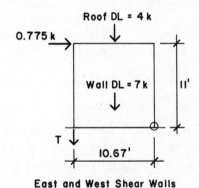

East and West Shear Walls

FIGURE 12.10 Building One: loading conditions for the shear walls.

tension in the edge framing members). The end reaction forces of 7.75 kips each become the lateral loads to the end walls and are distributed to the individual elements of the walls in proportion to their relative stiffnesses. Determined thusly, the loadings for the individual wall piers on all sides of the building are shown in Fig. 12.10. This assumes the individual pier form of action (Fig. 12.6a) and the pattern of walls shown in Fig. 12.1d and e.

For many of the available systems and components used for buildings such as Building One, rated capacities are established by codes or by the manufacturers of products. "Design" thus becomes largely a matter of matching load requirements, as determined by a loading investigation, to capacities of available elements. This is done most easily with a computer-aided process, but can also be done quite readily with available handbooks and manufacturer's brochures for very common systems.

12.4   BUILDING ONE: ALTERNATIVE CONSTRUCTION

Building One could be constructed with a great variety of materials and systems, without major alteration of the basic plan or the general exterior appearance of the building. A common alternative would be one using a frame structure of wood or steel with an exterior of masonry veneer. Figure 12.11 shows a possibility for the development of a system employing walls of precast concrete. Referring to the building plan in Fig. 12.1a, the north and south walls would be developed with individual 16 ft 8 in.-wide panels, each with one window.

The precast panel wall system could be developed to function in a variety of ways, such as

1. Fully structural units, resisting both gravity loading (as bearing walls) and lateral loading (as shear walls).
2. Shear walls only, with gravity loads supported on a separate frame, to which the panels are attached. (An alternative to this would be to cast concrete columns at the vertical joints between the panels, producing a separate frame and connecting the panels in a single operation.)
3. So-called architectural precast elements, serving no structural functions beyond their own needs.

The system shown in Fig. 12.11 would most likely be achieved by the tilt-up method, discussed in Sec. 3.10 and 10.6. While subject to much "customizing" in terms of finishes, dimensions, and some details, tilt-up systems are usually produced by companies that combine the activities of engineering design, unit production, and installation. They have the necessary expertise, experience, facilities, general "know-how," and most likely some patented devices and equipment for doing the work. General information about tilt-up construction is available from the ACI,

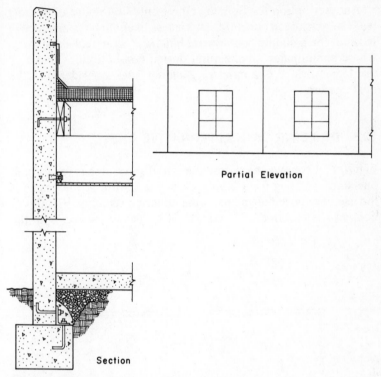

Partial Elevation

Section

FIGURE 12.11 Building One: alternative construction with sitecast, tilt-up concrete walls.

PCA, and PCI publications for those who wish to utilize the system. However, the work of any local producers should be investigated for individual building design projects.

Ordinary masonry or sitecast (cast-in-place) concrete can be made to achieve a considerable variety of building plans and forms. The use of the tilt-up structure—or just about any precast concrete system—generally requires considerable discipline in the form of planning modules and repetitive dimensioning. Suppliers of precast elements will bend over backwards to accommodate designer's requests, but the logical use of the systems indicates a need for some order and simplicity of planning.

Another option for Building One would be the use of sitecast walls of reinforced concrete. This is less likely to be economically feasible for a simple commercial building, but might be justified where some major architectural design considerations make its use logical. Such is a case for Building Two, considered in the next section.

12.5 BUILDING TWO: ALTERNATIVE ONE

Figure 12.12 shows the general form and some details for a small, one-story building for a library. A major feature of the design is the use of exposed elements of the concrete structure. Solid portions of the exterior walls consist of sitecast concrete, with the

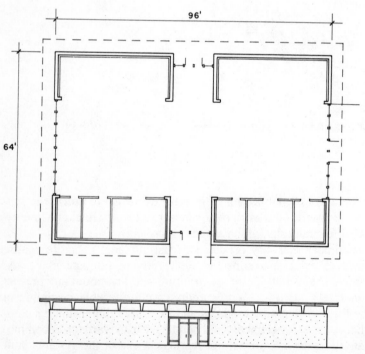

FIGURE 12.12 Building Two: Alternative One.

exterior surfaces exposed to view. The clear-span roof structure uses large precast concrete units, with the undersides exposed to view. The T-shaped roof elements extend over the concrete bearing walls to form an overhanging roof edge on the long sides of the building. Placement of the precast T units at the building ends also results in an overhanging roof edge.

Although both the roof and walls are concrete, the processes for their production are quite different, so a matching of the concrete materials and finishes is not possible. It would probably be best to develop the finishes in deliberately contrasting ways in order to further differentiate the two materials. The undersides of the T units may simply be cleaned and painted, while the exposed surfaces of the walls might receive a sandblasting or other treatment to produce a roughened surface, exposing the coarse aggregate materials.

The walls may be untreated for some utility buildings, but would most likely have some enhancement of the interior surfaces with an applied finish. For cold climates, it would also be necessary to use some insulation, as the solid concrete wall is not very insulative. One possible solution is to use a furred-out interior surface with gypsum drywall, creating a void space for insulation batts.

The T-shaped roof units are commercial products and would be supplied by a manufacturer who may also perform the site installation. Structural design of the units would be done by the manufacturer's engineers or consultants. Some dimensions and special details for the units would be adjusted to the specific usage situation, but the basic unit shapes would be variations on standard forms that the manufacturer produces. For the units shown here, a typical variable would be the actual unit width and depth, although the basic profile remains the same. Special steel elements would be cast into the units to permit attachment to supports, attachment of the adjacent units to each other, and possibly the attachment of some architectural elements for the windows, roof edge details, and so on.

The roof units deliver a considerable concentrated load to the tops of the concrete walls. A primary design investigation for the wall would be the effect of this loading and the required wall thickness that would be required. An investigation for a similar loading condition is done as the example problem in Sec. 10.2. In that problem it was found that a minimum wall thickness for the bearing condition was 6 in. However, the wall must also sustain lateral bending due to wind pressures or seismic effects. In addition, it must be poured in forms, and the 6-in. thickness is probably not feasible for the 12-ft-high pour indicated. A minimum thickness of 8 to 10 in. is probably more practical.

The combination of the very heavy roof structure and the heavy concrete walls will result in a considerable load to the wall footings. These footings may be several feet wide and would require some transverse reinforcement, but could still be selected from Table 9.3. Settlement of the footings for the bearing walls should be dealt with very conservatively, as the heavy walls will have a tendency to settle more than lighter portions of the building's interior construction. It is best to allow for some differential movements in the details of the construction, if possible, but really exaggerated differences in settlement are likely to cause distress for elements attached to the walls.

The flange portions of the T units may be made to act as a horizontal diaphragm by attaching adjacent units. This is usually done by welding of steel elements cast into the edges of the units at frequent intervals. The exterior concrete walls, with their turned ends in plan, constitute a very effective lateral bracing system in both directions of the building plan. Overturn is probably not critical, due to the weight of the construction. The major concern here would be for proper transfer of lateral forces from the T units to the walls at the seat connections for the units.

For the exposed wall a design concern is for the visibility of forming elements. The walls will most likely be formed with panels constructed with 4 ft by 8 ft sheets of plywood and lumber framing. The two opposed wall forms must be tied through the void to keep them from spreading under the hydraulic pressure of the wet concrete. The ties and form joints are very hard to hide on the exposed surface. The best design solution for this is to express the joints and ties deliberately by exaggerating them in a carefully controlled pattern. Figure 12.13 shows a partial elevation of the wall with a layout of form joints and ties that is coordinated with the module of the roof units. Of course, this kind of

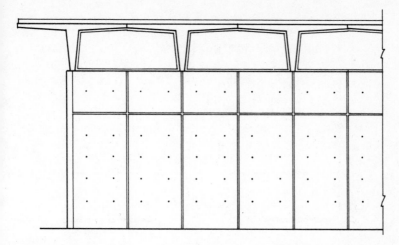

FIGURE 12.13   Building Two: exterior wall surface, developed to express forming joints and ties.

detail should be worked out in cooperation with the contractor for the concrete work.

With the absence of a suspended ceiling, it may be necessary to attach directly or suspend various elements from the T units for lighting, HVAC services, fire sprinklers, and so on. Provision for some attachments may be made by casting a continuous metal element into the bottom of the T stem. Standard hardware elements are available for this.

Many variations for both the wall and roof structures are possible for this building. The following sections illustrate two possible alternatives.

12.6   BUILDING TWO: ALTERNATIVE TWO

Figure 12.14 shows an alternative structure for Building Two, using a waffle slab system for the roof. Two interior columns are used to define a six-bay system for the waffle, with edges supported by a series of columns in the walls. The walls are formed

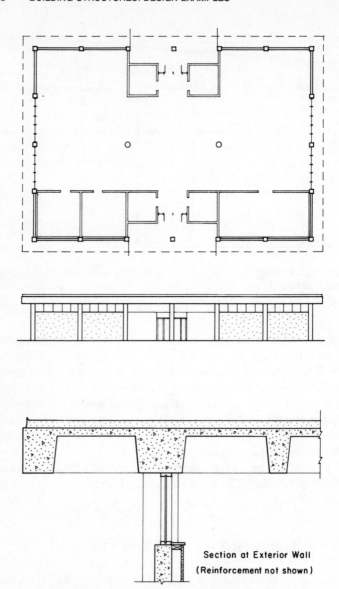

FIGURE 12.14 Building Two: Alternative Two.

with precast, tilt-up units, with the columns cast in place to connect the precast panels.

The waffle is formed with ordinary modular units, available from suppliers. The 30-in.-square "pans" are spaced 36 in. on center, forming rib joists with 6-in.-wide bottoms between the coffered voids. Wider ribs can be formed where necessary, such as that shown on the line of the columns above the walls.

Referring to the plan in Fig. 12.14, it may be observed that the waffle system is supported on the building perimeter by columns at 16 ft on center. A cantilevered overhang is developed with one unit of the waffle outside the columns, as shown in the section detail. The partial plan in Fig. 12.15 shows a reflected view of the waffle layout (like looking in a mirror on the floor). It may be

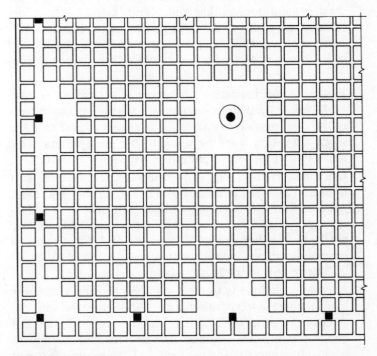

FIGURE 12.15 Building Two: reflected plan, showing the layout of the waffle system.

observed that there is a widened strip along the line of the perimeter columns and a solid portion around the interior column. The widened strip at the exterior columns facilitates the column-to-waffle joint and the window framing and constitutes a virtual continuous support at the exterior wall.

The solid portion around the interior column is the typical form of response to the concentration of shear and bending at the point of support. (See discussions in Secs. 7.5 and 7.6.) Although this is a standard structural detail for the waffle construction, it is not the only way to achieve the reinforcement of the concrete at this location. Some alternatives for this detail are shown in Fig. 12.16, consisting of the following:

1. Figure 12.16a indicates a shift of the waffle layout to have a void/coffer rather than a rib on the column line, permitting a smaller solid portion than that shown in Fig. 12.15. This reduces the width of the solid portion below the size ordinarily recommended, so the shear and bending would need to be investigated to demonstrate that this solution is adequate.

2. Figure 12.16b indicates a retaining of the layout in Fig. 12.15, except that 20-in.-square pans are used in the corners of the solid portion to reduce its mass slightly and ease the visual transition from the waffle to the solid portion. This tends to pull the waffle pattern into the solid portion.

3. In a similar fashion, Fig. 12.16c shows the use of 20-in. pans to extend the solid portion shown in Fig. 12.16a. The 20-in. pans are the other standard-size pan, ordinarily used with 4-in.-wide ribs to form a 24-in. modular system. Pans are also available in other special dimensions to facilitate special edge and corner conditions (10 in. square and 10 by 20; 15 in. square and 15 by 30, for example).

4. Another technique is to strengthen the resistance to shear by widening the top of the column, also slightly reducing the span in the process. This is a trick developed by ancient builders with stone, to allow the weak stone lintels to work with wider-spaced columns. Figure 12.16d shows the use of a spread-formed capital on the column.

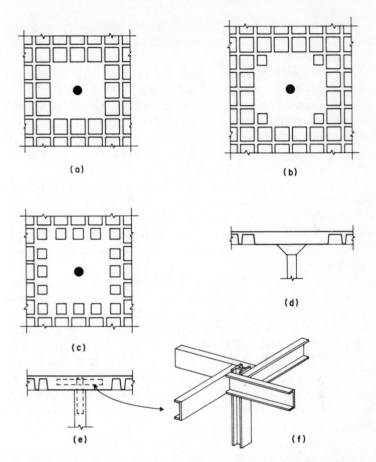

FIGURE 12.16 Building Two: alternative details at the interior columns.

5. Another way of achieving the column capital effect, where the capital as shown in Fig. 12.16 may not be desired, is to cast a steel element into the waffle and column that assists with both the shear transfer and development of bending moment. Figure 12.16e shows the installation of such a device, with Fig. 12.16f indicating its form as made up from rolled steel elements welded together.

These and other variations are possible considerations for either improving the structural character of the joint or improving its visible form.

The details shown so far have indicated the use of a round concrete interior column. Other forms are possible for this column, including a vertically tapered profile that could achieve the spread top while maintaining a relatively small size at the floor level (as in Frank Lloyd Wright's Johnson Wax Office Building). A steel column could also be used, possibly with a head detail similar to that shown in Fig. 12.16*f*.

The exterior column would be detailed and constructed as ordinary reinforced columns, as discussed in Chapter 9. They will constitute a rigid frame with the widened waffle rib, and their reinforcement would be bent at the top to develop some moment between the columns and the continuous beam. (See the discussion of Building Three.)

The exterior columns and beams could be developed to resist lateral loads, but for the construction as shown with the precast wall panels, the rigidity of the walls and the connections between the walls and columns will result in the panels working as a shear wall system. Because of the extent of the walls, the unit stress developed for the shear wall actions will be quite low. One option for the building would be to use a nonstructural curtain wall, in which case the exterior columns and the continuous beams over them would be developed as a rigid frame for lateral load resistance.

12.7 BUILDING TWO: ALTERNATIVE THREE

Figure 12.17 shows another possibility for the construction of Building Two, replacing the cast-in-place walls shown in Fig. 12.12 with walls of reinforced concrete masonry. In this system the concentrated loads from the T units could be resisted by internal reinforcement of the walls or by developing pilaster columns, as shown in the details.

The masonry walls will function essentially the same as the concrete walls, and will be designed for the gravity and lateral loads in the same manner. A concern for this structure is the

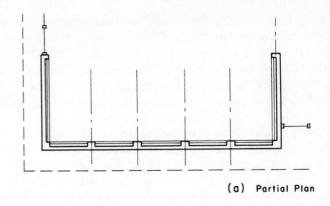

(a) Partial Plan

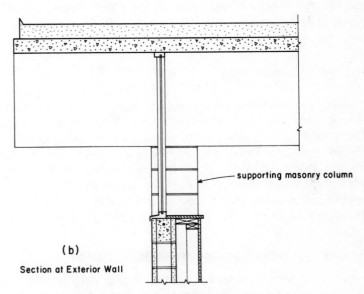

supporting masonry column

(b)

Section at Exterior Wall

FIGURE 12.17 Building Two: Alternative Three.

development of a modular system for plan and vertical layouts so that full masonry units can be used. The typical horizontal unit for this is 8 in., representing one-half of the typical 16-in.-long units. This is also a common vertical unit, although 6- and 4-in.-high units can also be obtained.

A variation for this system would be to use some cast-in-place concrete elements in combination with the masonry. A cast beam could be used along the tops of the walls as an architectural trim element and a means for distributing the concentrated loads from the T units to bearing on the masonry. It would also be possible to use cast concrete columns, in a manner similar to that in Sec. 12.6.

Masonry construction could also be used in combination with the waffle roof system described in Sec. 12.6. In all, there are a considerable number of alternatives for this building, if one considers the range of possible considerations for the roof and wall constructions with various combinations of masonry, sitecast, and precast concrete, and both conventionally reinforced and prestressed concrete.

12.8 BUILDING THREE

Building Three is a three-story office building, designed for speculative rental. As with Buildings One and Two, there are many alternatives for the construction of such a building, although in any given place at any given time, the basic construction of such buildings will most likely vary little from a limited set of choices. We will show here some of the work for the design of two alternatives for the construction: one using structural masonry walls and the other using a complete cast-in-place concrete structure.

General Considerations for the Building Design

Figure 12.18 presents a plan of the upper floor and a full building section for Building Three. We assume that a fundamental requirement for the building is the provision of a significant amount of exterior window surface and the avoidance of long expanses of unbroken solid wall surface. Another assumption is that the building is freestanding on the site so that all sides have a clear view. Most designers would also prefer that the space available for rental be as free as possible of permanent construction, permitting maximum flexibility in rearrangement for successive tenants.

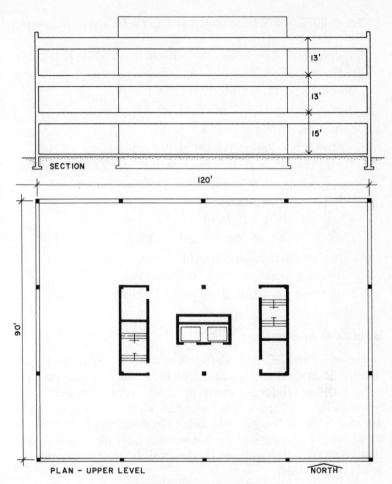

FIGURE 12.18 Building Three.

The latter translates to a general desire to eliminate permanent structural elements (columns or bearing walls) from the rental space—preferring them at the building perimeter and the location of permanent plan elements such as stairs, elevators, restrooms, and duct shafts.

The following will be assumed as criteria for the design work:

Building code: 1988 edition of Uniform Building Code (Ref. 4).
Live Loads:
   Roof: Table 11.2 (UBC Table 23-C)
   Floors: Table 11.3 (UBC Table 23-A)
   Office areas: 50 psf [2.39 kPa].
   Corridor and lobby: 100 psf [4.79 kPa].
   Partitions: 20 psf (UBC minimum per Sec. 2304) [0.96 kPa].
Wind: map speed, 80 mph [129 km/h]; exposure B.
Assumed construction loads:
   Floor finish: 5 psf [0.24 kPa].
   Ceilings, lights, ducts: 15 psf [0.72 kPa].
   Walls (average surface weight):
     Interior, permanent: 10 psf [0.48 kPa].
     Exterior curtain wall: 15 psf [0.72 kPa].

Structural Alternatives

The plan as shown, with 30-ft square bays and a general open interior, is an ideal arrangement for a beam and column system in either steel or reinforced concrete. Other types of systems may be made more effective if some modifications of the basic plans are made. These changes may affect the planning of the building core, the plan dimensions for the column locations, the articulation of the exterior wall, or the vertical distances between the levels of the building.

The general form and basic type of the structural system must relate to both the gravity and lateral force problems. Considerations for gravity require the development of the horizontal spanning systems for the roof and floors and the arrangement of the vertical elements (walls and columns) that provide support for the spanning structure. Vertical elements should be stacked, thus requiring coordinating the plans of the various levels.

The most common choices for the lateral bracing system would be the following (see Fig. 12.19):

1. *Core Shear Wall System* (Fig. 12.19*a*). This consists of using solid walls to produce a very rigid central core. The rest of the structure leans on this rigid interior portion, and the roof and floor constructions outside the core, as well as the exterior walls, are free from concerns for lateral forces as far as the structure as a whole is concerned.

2. *Truss-Braced Core.* This is similar in nature to the shear-wall-braced core, and the planning considerations would be essentially similar. The solid walls would be replaced by bays of trussed framing (in vertical bents) using various possible patterns for the truss elements.

3. *Perimeter Shear Walls* (Fig. 12.19*b*). This in essence makes the building into a tubelike structure. Because doors and windows must pierce the exterior, the peripheral shear walls usually consist of linked sets of individual walls (sometimes called piers).

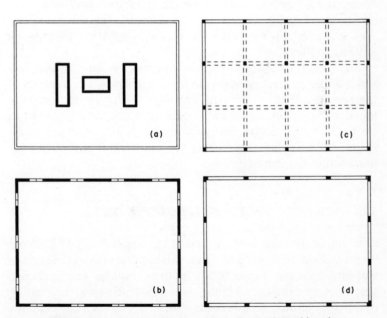

FIGURE 12.19 Building Three: options for the lateral bracing.

4. *Mixed Exterior and Interior Shear Walls.* This is essentially a combination of the core and peripheral systems.

5. *Full Rigid-Frame System* (Fig. 12.19c). This is produced by using the vertical planes of columns and beams in each direction as a series of rigid bents. For this building there would thus be four bents for bracing in one direction and five for bracing in the other direction. This requires that the beam-to-column connections be moment resistive.

6. *Perimeter Rigid-Frame System* (Fig. 12.19d). This consists of using only the columns and beams in the exterior walls, resulting in only two bracing bents in each direction.

In the right circumstances any of these systems may be acceptable. Each has advantages and disadvantages from both structural design and architectural planning points of view. The core-braced schemes were popular in the past, especially for buildings in which wind was the major concern. The core system allows for the greatest freedom in planning the exterior walls, which are obviously of major concern to the architect. The perimeter system, however, produces the most torsionally stiff building—an advantage for seismic resistance.

The rigid-frame schemes permit the free planning of the interior and the greatest openness in the wall planes. The integrity of the bents must be maintained, however, which restricts column locations and planning of stairs, elevators, and duct shafts so as not to interrupt any of the column-like beams. If designed for lateral forces, columns are likely to be large, and thus offer more intrusion in the building plan.

12.9 BUILDING THREE: ALTERNATIVE ONE

A structural framing plan for one of the upper floors of Building Three is shown in Fig. 12.20. The plan indicates the use of bearing walls as the major supports for the floor framing. The walls will also constitute the lateral bracing system, with some combination of a perimeter and core-braced system. There are many options for the floor framing, depending on fire code requirements and the local competitive pricing of suppliers. For the office building,

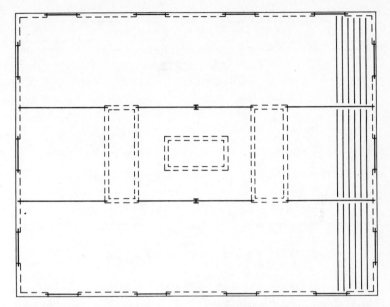

FIGURE 12.20 Building Three: framing layout plan for Alternative One; upper floor.

there are also many detailed concerns for incorporation of elements for wiring, piping, heat and cooling, ventilation, fire sprinklers, and lighting. We will assume that the various considerations can be met with a system consisting of a plywood deck, light nailable joists or trusses, and steel beams, as shown in Fig. 12.21.

Figure 12.22 shows the general construction of the exterior walls, indicating the use of reinforced concrete masonry. An exterior insulation system with applied finish is used on the outside surface and furring strips with gypsum drywall on the inside. The remainder of the discussion for this alternative deals with the design of the structural masonry walls.

Design for gravity loads is relatively simple. There are two general concerns: first for the general bearing in the walls and second for concentrated loads from supported beams. Vertical gravity loads in the walls are greatest in the first story (ground

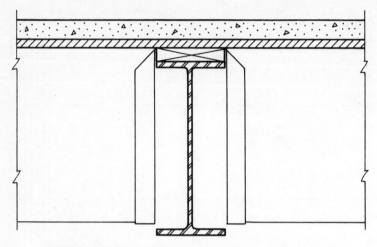

FIGURE 12.21 Building Three: detail of the floor structure.

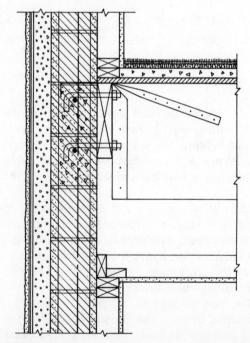

FIGURE 12.22 Building Three: detail of the wall construction.

level), so the masonry construction must facilitate this loading condition. For the reinforced concrete masonry there is a minimum construction that satisfies code requirements, typically requiring that at least every 4 ft of plan length of the wall have one concrete-filled, steel-reinforced void space. As the voids are typically 8 in. on center, this means at least every sixth void will be reinforced. From that minimum, it is possible to upgrade the wall strength by filling additional voids, and possibly by using more than the specified minimum reinforcement.

It is also possible to vary the wall thickness, as is more commonly done with *unreinforced* construction. However, for the three-story building, it is possible—and probably more feasible—to use a single thickness wall for all stories, with use of the minimum construction at the top and progressive upgrading with added reinforcement in lower stories.

As the framing plan in Fig. 12.20 shows, the interior steel beams and the lintels in the exterior walls are supported at ends or corners of the walls. These will automatically be locations of reinforced voids in this construction, and can most likely be developed to provide the necessary concentrated strengths. However, it is also possible to develop enlarged masonry elements (as pilasters) at these locations, if they are structurally required.

Design for Lateral Forces

The masonry walls must also develop resistance to lateral loadings, in combination with the horizontal roof and floor diaphragms. The final critical design of the walls must consider the combined effects of gravity and lateral loads. We first consider the effects of wind, using the criteria given previously and the requirements of the 1988 edition of the UBC (Ref. 4).

It is quite common, when designing for both wind and seismic forces, to have some parts of the structure designed for wind and others for seismic effects. In fact, what is necessary is to analyze for both effects and to design each element of the structure for the condition that produces the greater effect. Thus the shear walls may be designed for seismic effects, the exterior walls and window glazing for wind, and so on.

For wind it is necessary to establish the design wind pressure, defined by the code as

$$p = C_e C_q q_s I$$

where C_e is a combined factor, including concerns for the height above grade, exposure conditions, and gusts. From UBC Table 23-G, assuming exposure B:

$$C_e = 0.7 \text{ from 0 to 20 ft above grade}$$
$$= 0.8 \text{ from 20 to 40 ft}$$
$$= 1.0 \text{ from 40 to 60 ft}$$

and C_q is the pressure coefficient. Using the projected are method (method 2) we find from UBC Table 23-H the following.

For vertical projected area:

$$C_q = 1.3 \text{ up to 40 ft above grade}$$
$$= 1.4 \text{ over 40 ft}$$

For horizontal projected area (roof surface):

$$C_q = 0.7 \text{ upward}$$

The symbol q_s is the wind stagnation pressure at the standard measuring height of 30 ft. From UBC Table 23-F the q_s value for a speed of 80 mph is 17 psf.

TABLE 12.1 Design Wind Pressures for Building Two

Height above Average Level of Adjoining Ground (ft)	C_e	C_q	Pressure.[a] p (psf)
0–20	0.7	1.3	15.47
20–40	0.8	1.3	17.68
40–60	1.0	1.4	23.80

[a] Horizontally directed pressure on vertical projected area: $p = C_e \times C_q \times 17$ psf.

For the importance factor I (UBC TAble 23-K) we use a value of 1.0.

Table 12.1 summarizes the foregoing data for the determination of the wind pressures at the various height zones for Building Two. For the analysis of the horizontal wind effect on the building, the wind pressures are applied and translated into edge loadings for the horizontal diaphragms (roof and floors) as shown in Fig. 12.23. Note that we have rounded off the wind pressures from Table 12.1 in Fig. 12.23.

Figure 12.24a shows a plan of the building with an indication of the masonry walls that offer potential as shear walls for resistance to north-south lateral force. The numbers on the plan are the approximate plan lengths of the walls. Note that although the core construction actually produces vertical tubular-shaped elements, we have considered only the walls parallel to the load direction. The walls shown in Fig. 12.24a will share the total wind load delivered by the diaphragms at the roof, third-floor, and second-floor levels (H_1, H_2, and H_3, respectively, as shown in Fig. 12.23.) Assuming the building to be a total of 122-ft wide in

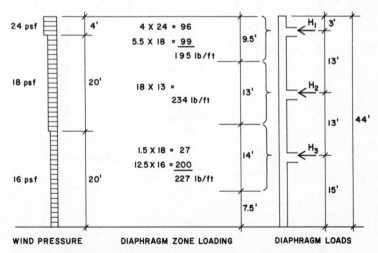

FIGURE 12.23 Building Three: investigation for lateral load.

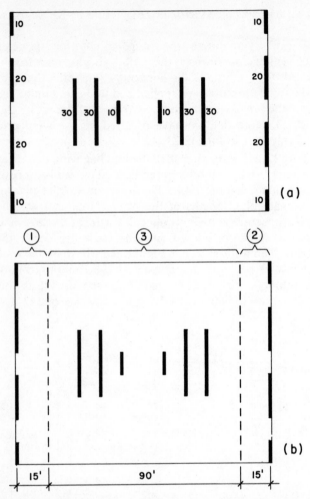

FIGURE 12.24 Building Three: considerations for the load distribution to the north–south shear walls.

the east-west direction, the forces at the three levels are

$$H_1 = 195 \times 122 = 23{,}790 \text{ lb } [106 \text{ kN}]$$

$$H_2 = 234 \times 122 = 28{,}548 \text{ lb } [127 \text{ kN}]$$

$$H_3 = 227 \times 122 = 27{,}694 \text{ lb } [123 \text{ kN}]$$

and the total wind force at the base of the shear walls is the sum of these loads, or 80,032 lb [356 kN].

Although the distribution of shared load to masonry walls is usually done on the basis of a more sophisticated analysis for relative stiffness, if we assume for the moment that the walls are stiff in proportion to their plan lengths (as is done with plywood walls), we may divide the maximum shear load at the base of the walls by the total of the wall plan lengths to obtain an approximate value for the maximum shear stress. Thus maximum shear:

$$v = \frac{80{,}032}{260}$$

$$= 308 \text{ lb/ft of wall length } [4.49 \text{ kN/m}]$$

This is quite a low force for a reinforced masonry wall, which tells us that if wind alone is of concern we will have considerable overkill in terms of total shear walls.

The shear stresses will not be equal in all walls because the forces will not be distributed evenly between the walls. We visualize the distribution of the total lateral force to the walls by considering two extreme cases regarding the stiffness of the horizontal diaphragms (roof and floor decks). The primary concern is for the functioning of the horizontal diaphragms. First, if these are considered to be infinitely stiff, then the distribution to the individual walls will be strictly in terms of their relative stiffness or deflection. Second, if the horizontal diaphragms are considered to be quite flexible (in their diaphragm spanning actions), then the distribution to the shear walls will be on a peripheral basis.

Figure 12.24b shows the building plan with the north-south shear walls and a breakdown of peripheral distribution assuming the flexible horizontal diaphragm. On this basis, the end shear walls each carry one eighth of the total shear and the core walls carry three fourths of the shear. In this approach the next step would be to consider the relative stiffness of the group of walls in each of the zones and to distribute forces to the individual walls.

In truth, the nature of the diaphragms is most likely somewhere between the two extremes described (just as most structural connections are neither pinned nor fully fixed, but actually partially fixed). It is thus not uncommon in practice for designers to investigate both conditions and to incorporate data from both analyses into their designs.

For either approach it is necessary to consider the relative stiffness of the walls of various plan length. For a description of this procedure, as well as a discussion of response to seismic loads, the reader is referred to Ref. 13 or 19.

12.10 BUILDING THREE: ALTERNATIVE TWO

A structural framing plan for the upper floors in Building Three is presented in Fig. 12.25, where the use of a cast-in-placed slab and beam system of reinforced concrete is indicated. Support for the spanning structure is provided by concrete columns. The system for lateral load resistance is that shown in Fig. 12.19d, which utilizes the exterior columns and spandrel beams as rigid-frame bents. This is a highly indeterminate structure for both gravity and lateral force design, and its precise engineering design would undoubtedly be done with a computer-aided system. We will discuss the major design considerations and illustrate the use of some simplified techniques for an approximate analysis and design of the structure.

Design of the Slab-and-Beam Floor Structure

As shown in Fig. 12.25, the basic floor-framing system consists of a series of beams at 10-ft centers that support a continuous, one-way spanning slab and are supported by column-line girders or directly by the columns. We will discuss the design of three elements of this system: the continuous slab, the four-span beam, and the three-span spandrel girder.

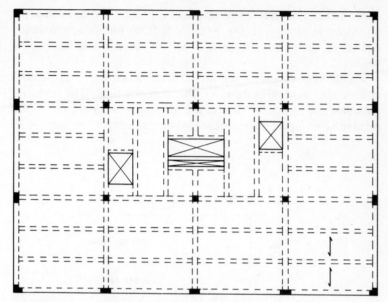

FIGURE 12.25 Building Three: framing layout plan for Alternative Two; upper floor.

The design conditions for slab, beam, and girder are indicated in Fig. 12.26. Shown on the diagrams are the positive and negative moment coefficients as given in Chapter 8 of the ACI Code (Ref. 1). Use of these coefficients is quite reasonable for the design of the slab and beam. For the girder, however, the pres-

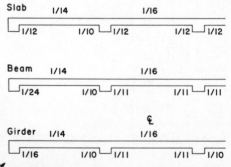

FIGURE 12.26 Building Three: approximate design moment coefficients for the slab-and-beam system.

ence of the concentrated loads makes the use of the coefficients improper according to the ACI Code. But for an approximate design of the girder, their use will produce some reasonable results.

Figure 12.27 shows a section of the exterior wall that demonstrates the general nature of the construction. The exterior columns and spandrel beams are exposed to view and would receive some special treatment for a higher degree of control of

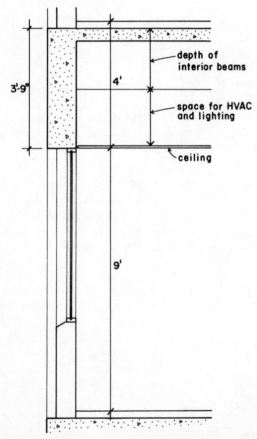

FIGURE 12.27 Building Three: section at the exterior wall, Alternative Two.

the finished concrete. The use of the full available depth of the spandrel beams results in a much stiffened frame on the building exterior, which partly justifies the choice of the perimeter bent system for lateral bracing.

The design of the continuous slab is presented as the example in Sec. 7.2. The use of the 5-in. slab is based on assumed minimum requirements for fire protection. If a thinner slab is possible, the 9-ft clear span would not require this thickness based on limiting bending or shear conditions or recommendations for deflection control. If the 5-in. slab is used, however, the result will tend to be a slab with a relatively low percentage of steel bar weight per square foot—a situation usually resulting in lower cost for the structure.

The unit loads used for the slab design in Sec. 7.2 are determined as follows:

Floor live load:
    100 psf (at the corridor) [4.79 kPa]
Floor dead load:
    Carpet and pad at 5 psf
    Ceiling, lights, and ducts at 15 psf
    2-in. lightweight concrete fill at 18 psf
    Assumed 5-in.-thick slab at 62 psf
        Total dead load: 100 psf [4.79 kPa]

Inspection of the framing plan in Fig. 12.25 reveals that there are a large number of different beams in the structure for the floor with regard to individual loadings and span conditions. Two general types are the beams that carry only uniformly distributed loads as opposed to those that also provide some support for other beams; the latter produce a load condition consisting of a combination of concentrated and distributed loading. We now consider the design of one of the uniformly loaded beams.

The beam that occurs most often in the plan is the one that carries a 10-ft-wide strip of the slab as a uniformly distributed loading, spanning between columns or supporting beams that are 30 ft on center. Assuming the supports to be approximately 12 in. wide, the beam has a clear span of 29 ft and a total load periphery of $29 \times 10 = 290$ ft^2. Using the UBC provisions for reduction of live load,

$$R = 10.08(A - 150)$$
$$= 0.08(290 - 150) = 11.2\%$$

We round this off to a 10% reduction, and, using the loads tabulated previously for the design of the slab, determine the beam loading as follows:

$0.90 \times 100 \times 10 = 900$ lb/ft   or   0.90 kip/ft [13.1 kN/m]

Slab and superimposed dead load:

$100 \times 10 = 1000$ lb/ft   or   1.0 kip/ft [14.6 kN/m]

The beam stem weight, estimating a size of 12×20 in. for the beam stem extending below the slab, is

$$\frac{12 \times 20}{144} \times 150 \text{ lb/ft}^3 = 250 \text{ lb} \quad \text{or} \quad 0.25 \text{ kip/ft } [3.65 \text{ kN/m}]$$

The total uniformly distributed load is thus

$$0.90 + 1.0 + 0.25 = 2.15 \text{ kip/ft } [31.35 \text{ kN/m}]$$

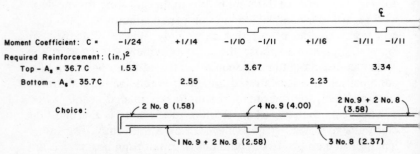

FIGURE 12.28 Building Three: design of the four-span beam.

Let us now consider the design of the four-span continuous beam that occurs in the bays on the north and south sides of the building and is supported by the north-south spanning column-line beams that we will refer to as the girders. The approximation factors for design moments for this beam are given in Fig. 12.26, and a summary of the design is presented in Fig. 12.28. Note that the design provides for tension reinforcing only, thus indicating that the beam dimensions are adequate to prevent a critical condition with regard to flexural stress in the concrete. Using the working stress method, the basis for this is as follows.

Maximum bending moment in the beam is

$$M = \frac{wL^2}{10}$$

$$= \frac{2.15(29)^2}{10}$$

$$= 181 \text{ kip-ft } [245 \text{ kN-m}]$$

Then, for a balanced section, using factors from Table 6.1,

$$\text{required } bd^2 = \frac{M}{R} = \frac{181 \times 12}{0.204}$$

$$= 10,647 \text{ in.}^3 \, [175 \times 10^6 \text{ mm}^3$$

If $b = 12$ in.,

$$d = \sqrt{\frac{10,647}{12}} = 29.8 \text{ in. } [757 \text{ mm}]$$

With minimum concrete cover of 1.5 in. on the bars, No. 3 U-stirrups, and moderate-sized flexural reinforcing, this d can be approximately attained with an overall depth of 33 in. (See Fig. 12.29.) This produces a beam stem that extends 28 in. below the slab, and is thus slightly heavier than that assumed previously. Based on this size, we will increase the design load to 2.25 kip/ft for the subsequent work.

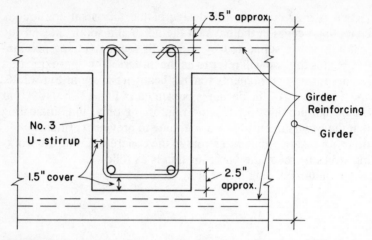

FIGURE 12.29 Building Three: reinforcement considerations for the beams and girders.

Before proceeding with the design of the flexural reinforcing, it is best to investigate the situation with regard to shear to make sure that the beam dimensions are adequate. Using the approximations given in Chapter 8 of the ACI Code, the maximum shear is considered to be 15% more than the simple span shear and to occur at the inside end of the exterior spans. We thus consider the following.

The maximum design shear force is

$$V = 1.15 \times \frac{wL}{2} = 1.15 \times \frac{2.25 \times 29}{2}$$

$$= 37.5 \text{ kips [167 kN]}$$

For the critical shear stress this may be reduced by the shear between the support and the distance of d from the support; thus

critical $V = 37.5 - \dfrac{29}{12} \times 2.25$

$$= 32.1 \text{ kips [143 kN]}$$

Using a d of 29 in., the critical shear stress is

$$v = \frac{V}{bd} = \frac{32,100}{29 \times 12} = 92 \text{ psi } [643 \text{ kPa}]$$

With the concrete strength of 3000 psi, this results in an excess shear stress of 32 psi that must be accounted for by the stirrups. The closest stirrup spacing would thus be

$$s = \frac{A_v f_s}{v' b} = \frac{0.22 \times 24,000}{32 \times 12}$$

$$= 13.75 \text{ in. } [348 \text{ mm}]$$

Because this results in quite a modest amount of shear reinforcing, the section may be considered to be adequate.

For the approximate design shown in Fig. 12.28, the required area of tension reinforcing at each section is determined as

$$A_s = \frac{M}{f_s j d} = \frac{C \times 2.25 \times (29)^2 \times 12}{24 \times 0.89 \times 29}$$

$$= 36.7 \, C$$

Based on the various assumptions and the computations we assume the beam section to be as shown in Fig. 12.29. For the beams the flexural reinforcing in the top that is required at the supports must pass either over or under the bars in the tops of the girders. Because the girders will carry heavier loadings, it is probably wise to give the girder bars the favored position (nearer the outside for greater value of d) and thus to assume the positions as indicated in Fig. 12.29.

At the beam midspans the maximum positive moments will be resisted by the combined beam and slab section acting as a T-section. For this condition we assume an approximate internal moment arm of $d - t/2$ and may approximate the required steel areas as

$$A_s = \frac{M}{f_s(d - t/2)}$$

$$= \frac{C \times 2.25 \times (29)^2 \times 12}{24 \times (29 - 2.5)} = 35.7C$$

The beams that occur on the column lines are involved in the lateral force resistance actions and are discussed later in relation to the design for lateral forces.

Inspection of the framing plan in Fig. 12.25 reveals that the girders on the north-south column lines carry the ends of the beams as concentrated loads at the third points of the girder spans. Let us consider the spandrel girder that occurs at the east and west sides of the building. This member carries the outer ends of the first beams in the four-span rows and in addition carries a uniformly distributed load consisting of its own weight and that of the supported exterior wall. The form of the girder and the wall was shown in Fig. 12.27. From the framing plan note that the exterior columns are widened in the plane of the wall. This is done to develop the perimeter bent system, as will be discussed later.

For the spandrel girder we determine the following:

Assumed clear span: 28 ft [8.53 m].

Floor load periphery, based on the carrying of two beams and half the beam span load is,

$$15 \times 20 = 300 \text{ ft}^2 \ [27.9 \text{ m}^2]$$

Note: This is approximately the same total load area as that carried by a single beam, so we will use the live load reduction of 10% as determined for the beam.

Loading from the beams:

Dead load: 1.35 kips/ft × 15 ft
        = 20.35 kips

Live load: 0.90 kip/ft × 15 ft
        = 13.50 kips

Total   33.85 kips, say 34 kips [151 kN]

Uniformly distributed load:

Spandrel beam weight: $\dfrac{12 \times 45}{144 \times 150} = 560$ lb/ft

Wall assumed at 25 psf: 25 × 9 = 225 lb/ft

Total = 785 lb/ft, say 0.8 kip/ft [11.7 kN/m]

For the uniformly distributed load approximate design moments may be found using the moment coefficients as was done for the slab and beam. Values for this procedure are given in Fig. 12.26. The ACI Code does not permit the use of this procedure for concentrated loads, but we may adapt some values for approximate design using moments for a beam with the third-point loading. Values of positive and negative moments for the third-point loading may be obtained from various references, including Refs. 3, 6, 11, 12, and 18.

Figure 12.30 presents a summary of the work for determining the design moments for the spandrel girder under gravity loading. Moment values are determined separately for the two types of load and then added for the total design moment.

We will not proceed further with the girder design at this point, for the effects of lateral loading must also be considered. The moments determined here for the gravity loading will be combined with those from the lateral loading in a following discussion.

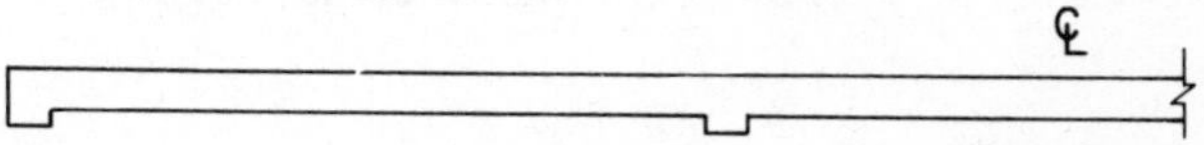

Moment due to distributed load: $M = C w L^2 = C \times 0.8 \times (28)^2 = 627\,C$

Coeff – C = –1/16	+1/14	–1/10	+1/16
M (k-ft) = –39.2	+44.8	–62.7	+39.2

Moment due to concentrated load: $M = C P L = C \times 34 \times 28 = 952\,C$

Coeff – C = –1/6	+2/9	–1/3	+1/6
M = –158.7	+211.6	–317.3	+158.7

Total gravity-induced moment:

M = –197.9	+256.4	–380	+197.9

FIGURE 12.30 Building Three: gravity loading conditions for the girders.

Design of the Concrete Columns

The four general cases for the columns are (Fig. 12.31):

1. The interior column carrying the primarily only axial gravity loads.
2. The intermediate exterior columns on the north and south sides carrying the ends of the interior girders and functioning as members of the peripheral bents for lateral resistance.
3. The intermediate exterior columns on the east and west sides carrying the ends of the column-line beams and functioning as members of the peripheral bents.
4. The corner columns carrying the ends of the spandrel beams and functioning as the end members in both peripheral bents.

Summations of the design loads for the columns may be done from the data given previously. As all columns will be subjected to combinations of axial compression and bending, these gravity

FIGURE 12.31 Building Three: framing arrangements for the columns.

Interior Column — Foundation to Roof

Size: 24" X 24" f'_c = 4 ksi f_y = 60 ksi

	Design service load (kips)	Reinforcing				Actual Capacity (kips) with e = 4"	
		Bars	P_g	Layout	Vertical arrangement	Ultimate	Service
Roof 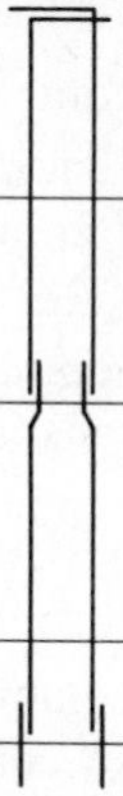							
13'	180	4 No.11	1.08 %	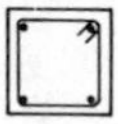		1116	446
3							
13'	337	4 No.11	1.08 %			1116	446
2							
15'	512	8 No.11	2.17 %	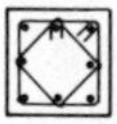		1254	502
1							
5'							
Foundation							

FIGURE 12.32 Building Three: design of the interior column for gravity only.

loads represent only the axial compression action. For the interior columns, the bending moments will be relatively low in comparison to the compression loads, and it is reasonable for a preliminary design to ignore bending effects and design for axial compression only. The usual minimum required eccentricity, as described in Sec. 8.3, will most likely provide for sufficient bending. On this basis, a trial design for one of the interior columns is shown in Fig. 12.32. Design loads were obtained from summations of dead loads plus reduced live loads. A single size of 24 in. square is used for all three stories, a common practice permitting the reuse of column forming for cost savings. The service load capacities indicated may be compared with values obtained from the graphs in Fig. 8.7. Economy is also generally obtained with the use of low percentages of reinforcement when bending moments are not a critical concern; the percentages shown in Fig. 12.32 are minimal, but smaller column sizes could be used if floor space and planning problems are of major concern.

The interior column occurs at the location of the stairs and rest rooms, and it is possible that some form alteration may be made to allow the columns to fit more smoothly into the wall planning. This will add cost to the column construction, but is relatively easily achieved, as shown in Fig. 8.8.

For the intermediate exterior column there are four actions to consider:

1. The vertical compression induced by gravity.
2. Bending moment induced by the interior framing that intersects the wall column; the columns are what provides the end moments shown in Figs. 12.28 and 12.30.
3. Bending moment in the plane of the wall induced by unbalanced conditions in the spandrel beams and girders.
4. Bending moments induced by the actions of the perimeter bents in resisting lateral loads.

For the corner column the situation is similar to that for the intermediate exterior column, that is, bending on both axes. The forms of the exterior columns as shown on the plan in Fig. 12.25 have been established in anticipation of the major effects described. Further discussion of these columns will be deferred, however, until after we have investigated the situations of lateral loading.

Design for Lateral Forces

The lateral force resisting systems for the concrete structure are shown in Fig. 12.33. For force in the east-west direction the resistive system consists of the horizontal roof and floor slabs and the exterior bents (columns and spandrel beams) on the north and south sides. For force in the north-south direction the system utilizes the bents on the east and west sides. Actually other elements of the structural frame will also resist lateral force, but by widening the columns and deepening the spandrel beams in these bents, an increased stiffness will be produced; the stiffer bents will then tend to offer the most resistance to lateral movements. We will design these stiffened bents for all the lateral force, ignoring the minor resistances offered by the other column and beam bents.

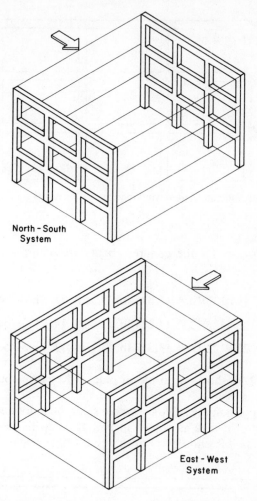

North – South
System

East - West
System

FIGURE 12.33   Building Three: the perimeter bent bracing system.

With the same building profile, the wind loads on this structure
will be the same as those determined for the masonry structure in
Sec. 12.9. As in the example in that section, we will illustrate the
design of the bents in only one direction, in this case the bents on
the east and west sides. From the data given in Fig. 12.23, the

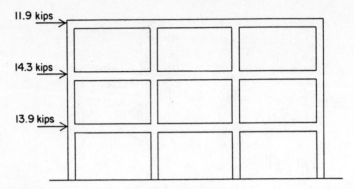

FIGURE 12.34 Building Three: wind loading for the north–south perimeter bents.

horizontal forces for the concrete bents are determined as follows:

$$H_1 = 195(122)/2 = 11,895 \text{ lb}, \qquad \text{say } 11.9 \text{ kips/bent}$$
$$H_2 = 234(122)/2 = 14,274 \text{ lb}, \qquad \text{say } 14.3 \text{ kips/bent}$$
$$H_3 = 227(122)/2 = 13,847 \text{ lb}, \qquad \text{say } 13.9 \text{ kips/bent}$$

Figure 12.34 shows a profile of one of the north-south bents with the bent loads.

For an approximate analysis we consider the individual stories of the bent to behave as shown in Fig. 12.35, with the columns developing an inflection point at their midheight points. Because

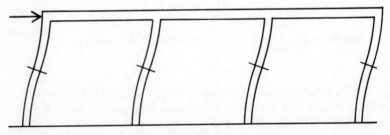

FIGURE 12.35 Building Three: assumed form of deformation of the bent columns.

the columns all move the same distance, the shear load in a single column may be assumed to be equal to the cantilever deflecting load and the individual shears to be proportionate to the stiffnesses of the columns. If the columns are all of equal stiffness in this case, the total load would be simply divided by 4. However, the end columns are slightly less restrained as there is a beam on only one side. We will assume the net stiffness of the end columns to be one-half that of the interior columns. Thus the shear force in the end columns will be one-sixth the load and that in the interior

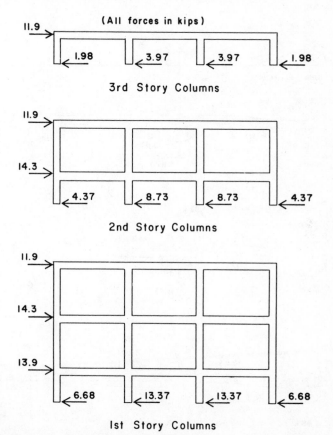

FIGURE 12.36 Building Three: investigation for column shear in the north–south bent.

columns one-third of the load. The column shears for each of the three stories is thus as shown in Fig. 12.36.

The column shear forces produce moments in the columns. With the column inflection points assumed at midheight, the moment produced by a single shear force is simply the product of the force and half the column height. These moments must be resisted by the end moments in the rigidly attached beams, and the actions are as shown in Fig. 12.37. These effects due to the lateral loads may now be combined with the previously determined effects of gravity loads for an approximate design of the columns and beams.

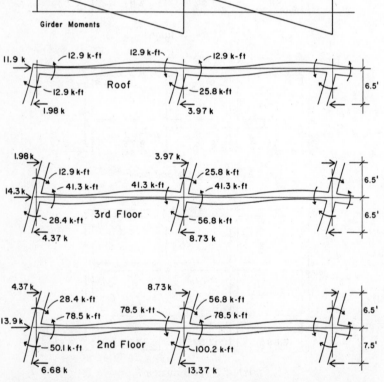

FIGURE 12.37 Building Three: investigation for column and girder bending moments in the north–south bent.

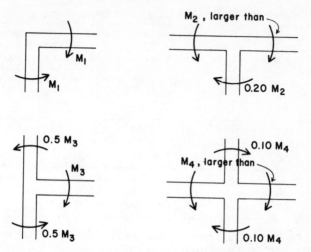

FIGURE 12.38 Assumptions for approximations of the distribution of bending moments in the bent columns due to gravity loads.

For the columns, we combine the axial compression forces with any gravity-induced moments and first determine that the load condition without lateral effects is not critical. We may then add the effects of the moments caused by lateral loading and investigate the combined loading condition, for which we may use the one-third increase in allowable stress. Gravity-induced beam moments are taken from Fig. 12.30 and are assumed to induce column moments as shown in Fig. 12.38. The summary of design conditions for the corner and interior column is shown in Table 12.2. The design values for axial load and moment and approximate sizes and reinforcing are shown in Fig. 12.39. Column sizes and reinforcing were obtained from the tables in the *CRSI Handbook* (Ref. 3) using concrete with $f'_c = 4$ ksi and grade 60 reinforcing.

The spandrel beams (or girders) must be designed for the combined shears and moments due to gravity and lateral effects. Using the values for gravity-induced moments from Fig. 12.30 and the values for lateral load moments from Fig. 12.37, the combined moment conditions are shown in Fig. 12.40. For design we must

	Intermediate Column					Corner Column				
	Axial Load	Moment	e	Column Dimensions	Reinforcement	Axial Load	Moment	e	Column Dimensions	Reinforcement
	(kips)	(kip-ft)	(in.)	(in.)	No. - Size	(kips)	(kip-ft)	(in.)	(in.)	No. - Size
Roof										
	90 X 3/4 = 68	85.8 X 3/4 = 64	11.3	20X28	6 - 9	55	120	35	20X24	6 - 10
3										
	179 X 3/4 = 134	95.8 X 3/4 = 72	6.5	20X28	6 - 9	117	100	13.6	20X24	6 - 10
2										
	277 X 3/4 = 208	139.2 X 3/4 = 105	6.1	20X28	6 - 10	176 X 3/4 = 132	150.1 X 3/4 = 113	10.3	20X24	6 - 10
1										

FIGURE 12.39 Building Three: design of the north–south bent columns for combined gravity and wind loads.

consider both the gravity only moment and the combined effect. For the combined effect we use three-fourths of the total combined values to reflect the allowable stress increase of one-third.

Figure 12.41 presents a summary of the design of the reinforcing for the spandrel beam at the third floor. If the construction

TABLE 12.2 Summary of Design Data for the Bent Columns

	Column	
	Intermediate	Corner
Axial gravity design load (kips)		
Third story	90	55
Second story	179	117
First story	277	176
Assumed gravity moment on bent axis (kip-ft) from figs. 12.30 and 12.38		
Third story	60	120
Second story	39	100
First story	39	100
Moment from lateral force (kip-ft) from Fig. 12.37		
Third story	25.8	12.9
Second story	56.8	28.4
First story	100.2	50.1

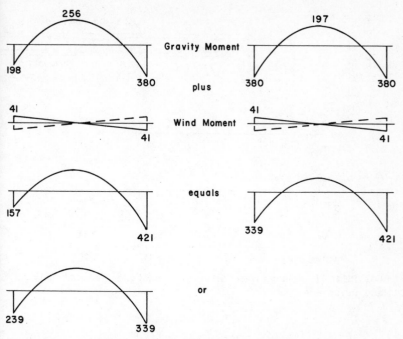

256

197

Gravity Moment

198

380

plus

380

380

41

41

Wind Moment

41

41

equals

157

339

421

421

or

239

339

FIGURE 12.40 Building Three: combined moments for the bent girder.

that was shown in Fig. 12.27 is retained with the exposed spandrel beams, the beam is quite deep. Its width should be approximately the same as that of the column, without producing too massive a section. The section shown is probably adequate, but several additional considerations must be made as will be discussed later.

For computation of the required steel areas we assume an effective depth approximately 40 in. and use

$$A_s = \frac{M}{f_s j d} = \frac{M(12)}{24(0.9)(40)} = 0.0139M$$

Because the beam is so deep, it is advisable to use some longitudinal reinforcing at an intermediate height in the section, especially on the exposed face.

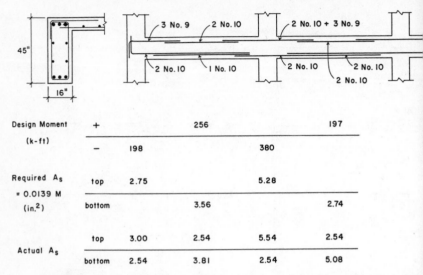

Design Moment				
+		256		197
(k-ft) −	198		380	

Required A$_s$					
= 0.0139 M	top	2.75		5.28	
(in.²)	bottom		3.56		2.74

Actual A$_s$					
	top	3.00	2.54	5.54	2.54
	bottom	2.54	3.81	2.54	5.08

FIGURE 12.41 Building Three: design of the bent girder for combined wind and gravity moments.

Shear design for the beams should also be done for the combined loading effects. The closed form for the shear reinforcing, as shown in Fig. 12.41, is used for considerations of torsion as well as the necessity for tying the compressive reinforcing.

With all of the approximations made, this should still be considered to be a very preliminary design for the beam. It should, however, be adequate for use in preliminary architectural studies and for sizing the members for a dynamic seismic analysis and a general analysis of the actions of the indeterminate structure.

12.11 BUILDING THREE: ALTERNATIVE THREE

Figure 12.42 shows a partial plan and some details for a concrete flat slab system for the roof and floor structures for Building Three. The general nature of this system is discussed in Sec. 7.6. Features of the system as shown in Fig. 12.42 include the following:

1. A general solid 10-in.-thick slab without beams in the major portions of the structure outside the plan core.

2. A thickened portion, called a *drop panel*, around the supporting columns, ordinarily extending one third of the span out from the columns on all sides. The thickness increase shown is one-half of the thickness of the slab. This is a typical average, but other thicknesses can be used.

3. A *column capital*, consisting of a truncated, inverted pyramidal form. This serves to slightly reduce the critical span and to reduce shear stresses in the drop panel.

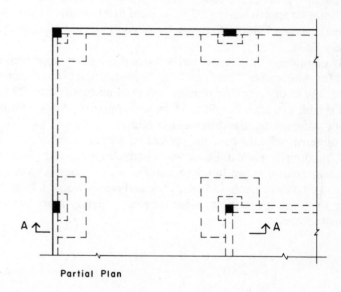

Partial Plan

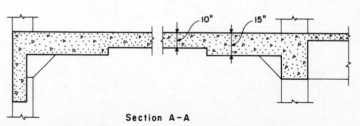

Section A-A

FIGURE 12.42 Building Three: Alternative Three.

4. Use of the same spandrel beam (as shown in Fig. 12.27) and slab and beam framing at the core as in Alternative Two.

The flat slab is generally more feasible when live loads are very high and there are a considerable number of continuous bays of the framing in both directions. The system here is marginally feasible, and might be justified on the basis of general building design considerations. Regarding the latter, a potential advantage to be had is the possibility of shortening the floor-to-floor and floor-to-roof vertical dimensions. As shown in Fig. 12.27, the total dimension provided between the top of the floor finish above and the underside of the ceiling below is 4 ft. If the typical interior beam size of approximately 32 in. is used in Alternative Two, the largest insulated duct that can be accommodated is approximately 12 in.

By comparison with Alternative Two, if the 10-in. solid slab is used for Alternative Three, and the same-size duct is accommodated, the floor-to-ceiling dimension could be reduced by 22 in. This is a bit extreme, but if the 15-in. drop panel thickness is used for the comparison, the difference is still as much as 17 in. Using that dimension for the roof and upper two floors, a total of 3.5 ft in total building height can be saved. The resultant savings in construction costs for the building exterior skin, stairs, elevators, vertical ducts and risers for the wiring and piping would add up to more than enough to offset what may be a slightly more expensive structure.

REFERENCES

1. *Building Code Requirements for Reinforced Concrete*, ACI 318-89, American Concrete Institute, Detroit. (Commonly called the ACI Code; currently published with accompanying commentary.)

2. *Building Code Requirements for Reinforced Concrete*, ACI 318-63, American Concrete Institute, Detroit. (The last edition of the ACI Code with a complete development of the working stress method, which is now given as the Alternate Design Method in Appendix A of the 1989 Code.)

3. *CRSI Handbook,* Concrete Reinforcing Institute, Schaumburg, IL, 1984.

4. *Uniform Building Code,* 1988 ed., International Conference of Building Officials, Whittier, CA.

5. *Reinforced Concrete Fundamentals*, 4th ed., P. Ferguson, Wiley, New York, 1979.

6. *Simplified Design: Reinforced Concrete Buildings of Moderate Size and Height,* Portland Cement Association, Skokie, IL, 1984.

7. *Notes on ACI 318-83 Building Code Requirements for Reinforced Concrete with Design Applications*, Portland Cement Association, Skokie, IL, 1984.

8. *ACI Detailing Manual—1988*, SP-66, American Concrete Institute, Detroit, 1988.

9. *Architectural Graphic Standards*, 8th ed., C. Ramsey and H. Sleeper, Wiley, New York, 1988.

10. *Minimum Design Loads for Buildings and Other Structures*, ANSI A58.1, American National Standards Institute, New York, 1982.

11. *Simplified Engineering for Architects and Builders*, 7th ed., H. Parker and J. Ambrose, Wiley, New York, 1989.

12. *Building Structures*, J. Ambrose, Wiley, New York, 1988.

13. *Simplified Design of Building Foundations*, 2nd ed., J. Ambrose, Wiley, New York, 1989.

14. *Simplified Building Design for Wind and Earthquake Forces*, 2nd ed., J. Ambrose and D. Vergun, Wiley, New York, 1990.

15. *Structural Details for Concrete Construction*, M. Newman, McGraw-Hill, New York, 1988.

16. *Structural Details for Masonry Construction*, M. Newman, McGraw-Hill, New York, 1988.

17. *Building Code Requirements for Masonry Structures* and *Specifications for Masonry Structures* (combined edition), ACI-ASCE 530-88 and 530.1-88, American Concrete Institute, Detroit, 1988.

18. *Concrete Masonry Handbook for Architects, Engineers, and Builders,* 7th ed., Portland Cement Association, Skokie, IL, 1985.

19. *Reinforced Masonry Design*, 2nd ed., R. Schneider and W. Dickey, Prentice-Hall, Englewood Cliffs, NJ, 1987.

20. *PCI Design Handbook: Precast and Prestressed Concrete*, 3rd ed., Prestressed Concrete Institute, Chicago, 1985.

21. *Tilt-Up Construction*, C-7, American Concrete Institute, Detroit, 1986.

22. *Building Code Requirements for Structural Plain Concrete*, ACI 318.1-89, American Concrete Institute, Detroit, 1989.

STUDY AIDS

The materials in this section are provided so that readers may measure their comprehension of the book presentations. It is recommended that upon completion of an individual chapter, the materials given here for that chapter be used as a review. Answers to the questions are provided at the end of this section.

Words and Terms

Using the text of the chapter indicated, together with the Index, review the meanings of the following words and terms.

Chapter 1

Concrete
Cement
Portland cement
Reinforced concrete
Sitecast concrete
Precast concrete

CMU
ACI
ACI Code
PCA
CRSI
PCI
NCMA
MIA
Working stress method
Service load
Strength method
Factored load
Alternative design method (ACI Code)

Chapter 2

Aggregate
High-early-strength cement
Air-entrained concrete
Admixture
Specific compressive strength, f'_c
Modulus of elasticity
Creep
Modulus ratio, n
Workability
Cover
Reinforcement
Prestressing
Pretensioning
Post-tensioning
Lightweight concrete
Insulating concrete
Fiber-reinforced concrete

Chapter 3

Initial set
Segregation
Slump test
Compression test
Shop drawings
Tilt-up wall
Reinforced masonry

Chapter 4

Construction joint
Control joint
Minimum reinforcement
Shrinkage and temperature reinforcement
Spacing requirements

Chapter 5

Continuous beam
Rigid frame
Moment-resistive connection
Free-body diagram
Deformed shape

Chapter 6

Balanced section
Under-reinforced section
Required strength (strength method)
Design strength (strength method)
Strength reduction factor
Rectangular stress block
T-beam

Stirrup
Development length
Standard hook
Lapped splice

Chapter 7

One-way slab and beam system
Two-way slab and beam system
Joist construction
Flat slab
Waffle construction
Composite construction

Chapter 8

Column interaction response
Tied column
Spiral column

Chapter 9

Shallow bearing foundation
Footing
Site survey
Allowable bearing pressure
Active lateral pressure
Passive lateral pressure
Presumptive soil properties
Deep foundations
Rectangular footing
Combined footing
Cantilever footing
Pedestal

Chapter 10

Bearing wall
Retaining wall
Shear wall
Freestanding wall
Grade wall (beam)

Chapter 11

Dead load
Building code
Live load
Live-load reduction
Lateral load
Seismic effect

GENERAL QUESTIONS

Note: Answers follow the last question.

Chapters 1 to 5

1. What is the primary structural limitation of concrete that generates the need for reinforcement?

2. What is the significance of having a well-graded range of sizes for the aggregate in concrete?

3. What is the primary factor that determines the unit density (weight) of structural concrete?

4. What is the purpose of the deformations on the surface of steel bars used for reinforcing concrete?

5. What property is most significant for the steel used for reinforcing bars for concrete?

6. During the curing period for concrete (after casting and before major usage), what significant controls should be exercised?

7. What are the significant concerns that establish cover requirements for reinforcements in concrete structures?

8. With regard to the assumption of stress conditions, what is the difference between the working stress and strength methods of design for reinforced concrete?

Chapter 6

1. What internal force development is represented by the stress wedge in working stress investigation of a reinforced concrete beam?

2. Other than spacing limits and bar diameters, what factors establish the maximum number of bars that can be placed in a single layer in a reinforced concrete beam?

3. Why is compressive stress in the concrete generally not critical for T-beam actions in sitecast construction?

4. In evaluating moment resistance for a doubly reinforced concrete beam, why is stress in the compressive reinforcement not always the same as that in the tensile reinforcement?

5. For shear actions in concrete beams, what acts together with the vertical stirrups to resist the diagonal tension stresses?

6. Stirrups are generally designed to resist what portion of the total shear force in a concrete beam?

7. Why is development length generally less critical for small-diameter reinforcing bars?

8. Why are longer development lengths required for bars of higher-grade steel?

9. In what form is the anchorage capacity of a standard hook expressed?

10. What is the basis for establishment of the length required for a lapped splice?

Chapter 7

1. What are the usual considerations for determination of slab thickness in a slab-and-beam system?

2. How is the minimum depth required for slabs and beams affected by span and support conditions?

3. What is the essential difference in structural action between concrete joist construction and waffle construction?

4. What structural improvements are achieved by the use of column capitals and drop panels in flat slab construction?

5. What is the essential function of the shear developers (welded steel studs) in composite construction with steel beams and a sitecast concrete slab?

Chapter 8

1. With regard to effects on the vertical reinforcement, what is the primary purpose of the ties in a tied column?

2. With a column subjected to a large bending moment, why is a slightly higher moment possible with the addition of a minor axial compression load?

3. What is the usual basis for limitation of the number of bars that can be placed in a spiral column?

Chapter 9

1. What are the principal engineering properties of soils that most affect building foundation design?

2. Other than practical forming considerations, what favors the use of a square plan form for column footings?

3. What is the purpose of the longitudinal reinforcement in a wall footing?

4. How does the use of a pedestal result in the possibility of a thinner column footing?

Chapter 10

1. When a concrete basement wall is reinforced with a single layer of bars, why are the bars located nearer the inside face of the wall?

2. What is the purpose of the projected portion beneath the footing of a cantilever retaining wall?

ANSWERS TO THE GENERAL QUESTIONS

Chapters 1 to 5

1. Low resistance to tensile stress, as generated primarily by bending and shear actions.

2. The aggregate, separately considered, will pack into the most dense mass, with smaller pieces filling the voids between larger ones.

3. The unit density of the coarse aggregate (gravel in ordinary concrete), which makes up the major portion of the concrete volume.

4. To develop a mechanical lock (grip) between the concrete and the bar surface.

5. The yield strength.

6. Temperature, moisture content, and lack of stress.

7. Size of the concrete member, fire resistance requirements, and exposure conditions (weather or soil).

8. The working stress method uses service (usage, working) level stress conditions. The strength method considers only failure conditions.

Chapter 6

1. Development of the internal compression force in the concrete that opposes the tension force in the steel to develop bending resistance.

2. Beam width, presence and size of stirrups, size of largest aggregate, and general code requirements—such as minimum spacing.

3. Typical sizes of slabs (thickness) and beams (primarily depth) result in excessive compression area for development of stress necessary to balance the tension in practical amounts of reinforcement.

4. For compatability of strain between the compressive reinforcement and the concrete around it, steel stress is often limited.

5. The horizontal, tensile reinforcing bars at the ends of the beam and the concrete section.

6. The shear in excess of that permitted to be resisted by the concrete.

7. Development occurs on the bar surfaces, to "develop" the strength of the bar in tensile stress on its cross-sectional area. The smaller the bar, the greater the ratio of surface to area and the more surface available in proportion to the bar tensile strength.

8. Required development relates to potential bar strength, which is greater if the allowable stress is higher.

9. In units of equivalent bar length.

10. The development length required for the bars being spliced.

Chapter 7

1. Span of the slab, design loading, fire resistance requirements, and maybe the T-beam action of the framing beams.

2. These form the basis for the recommended minimum thickness in the ACI Code.

3. Joists form a one-way span system; the waffle is a two-way span system.

4. Slab clear span is slightly reduced and bending and shear resistances of the slab are increased.

5. To make the steel beam and concrete slab work together as a single unit in resisting bending.

Chapter 8

1. To prevent the bars from buckling and bursting out sideways through the thin concrete cover.

2. Code-imposed design limitations result in columns having the yielding of the bars as the initial failure of the column in bending. Adding a small axial compression force prestresses the bars in compression, permitting them to take some additional tension and thus develop slightly more bending moment. Eventually, however, additional compression will result in a column failure action, reducing the reserve capacity for bending.

3. Spacing requirements for the bars.

Chapter 9

1. Size of solid particles, in-place (undisturbed) density and hardness, water content, penetration resistance, potential stability, presence of organic matter.

2. Ease and simplicity of placing the reinforcement.

3. To resist shrinkage stresses and maybe develop some beam spanning action over uneven soil conditions.

4. By reducing shear and bending in the footing.

Chapter 10

1. Tension develops on the inside as the wall spans vertically in opposing the soil pressure on the outside surface.

2. Enhancement of the resistance of the wall to horizontal sliding.

ANSWERS TO PROBLEMS

Chapter 5

5.8.A. $R = 10$ kips (up) and 110 kip-ft counterclockwise

5.8.B. $R = 5$ kips (up) and 24 kip-ft counterclockwise

5.8.C. $R = 6$ kips (to the left) and 72 kip-ft counterclockwise

5.8.D. Left $R = 4.5$ kips (up), right $R = 4.5$ kips (down and 12 kips to right

5.8.E. Left $R = 4.5$ kips (down) and 6 kips to left, right $R = 4.5$ kips (up) and 6 kips to left

Chapter 6

6.4.A. Width required to get bars into single layer is critical concern; least width: 16 in. with $h = 31$ in. and five No. 10 bars

6.4.B. From Problem 6.4.A, this is an under-reinforced section; find actual $k = 0.886\pm$, required $A_s = 5.08$ in.^{2}, use four No. 10 bars

6.5.A. Use $d = 11$ in., $A_s = 2.73$ in.^{2}
If $d = 16.5$ in., $A_s = 1.58$ in.^{2}

6.5.B. Use $d = 14$ in., $A_s = 4.35$ in.^{2}
If $d = 21$ in., $A_s = 2.74$ in.^{2}

6.5.C. Use $d = 14.5$ in., $A_s = 5.40$ in.^{2}
If $d = 22$ in., $A_s = 3.22$ in.^{2}

6.6.A. $M_R = 150$ kip-ft, use tension reinforcing of two No. 11 plus three No. 10, compressive reinforcing of three No. 9

6.6.B. Use tension reinforcing of five No. 9, compressive reinforcing of two No. 8

6.6.C. Use tension reinforcing of four No. 9, compressive reinforcing of two No. 8

6.7.A. 5.76 in.^{2}

6.8.A. Use 8-in.-thick slab, with No. 4 at 3 in., No. 5 at 5 in., No. 6 at 7 in., or No. 7 at 10 in.; use temperature reinforcing of No. 4 at 12

6.8.B. Use 8-in. slab (for deflection), No. 4 at a, No. 5 at 6, No. 6 at 9, No. 7 at 12 (approximately 15% less than by working stress)

6.9.A. Possible choice: 1 at 6 in., 8 at 13 in.

6.9.B. Possible choice: 1 at 6 in., 4 at 13 in.

6.10.A. $L_2 = 36$ in., less than 48 in. required, L_1 plus hook equals 29.1 in., less than 35 in. required

6.10.B. $L_2 = 42$ in., less than 68 in. required, L_1 plus hook equals 40.4 in., less than 48 in. required

6.10.C. $L_2 = 38$ in., less than 41 in. required, L_1 plus hook equals 32.4 in., greater than required 29 in.

6.10.D. $L_2 = 35$ in., just short of 36 in. required, L_1 plus hook = 31.4 in., greater than required of 25 in.

Chapter 7

7.2.A. Need 5 in. for bending moment, with all No. 5 bars spacings are 9 in. at outside end and second interior support, 7 in. at first interior support, 10 in. at first span center, 12 in. at second span center

Chapter 8

Note: Two answers are given when the point falls very close to a line in the graph.

8.8.A. Curve 8: 12-in. column, four No. 11

8.8.B. Curve 10: 16-in. column, four No. 10

8.8.C. Curve 13: 20-in. column, four No. 9

8.8.D. Curve 14: 20-in. column, eight No. 9 bars

8.8.E. Curve 16: 20-in. column, 12 No. 11 bars

8.9.A. Curve 8: 14-in. column, six No. 9, or curve 9: 16-in. column, four No. 7

8.9.B. Curve 14: 20-in. column, four No. 11

8.9.C. Curve 16: 20-in. column, eight No. 11, or curve 17: 24-in. column, four No. 10

8.9.D. Curve 18: 24-in. column, six No. 11

8.9.E. Curve 21: 30-in. column, six No. 10

Chapter 9

9.5.A. Possible choice: $w = 6$ ft 8 in., $h = 16$ in., seven No. 5 in long direction, No. 5 at 9-in. short direction

9.6.A. Possible choice: closest choice from Table 9.6, 9 ft square, $h = 21$ in., 10 No. 8 each way; computations should verify adequacy; other designs possible

9.8.A. (1) Footing could be 9 ft 9 in. square, 21 in. thick, with 10 No. 9 each way, except No. 11 column bars require 30 in. for development
(2) Could use 30 in. square by 30 in. pedestal, approximately same footing

Chapter 10

10.2.A. Factored load (P_u) is 63 kips, P_u/ϕ is 90 kips, critical bearing pressure is 1500 psi—less than allowable of 1785 psi, allowable capacity by column load formula is 356 kips (greater than P_u/ϕ), minimum reinforcement is No. 4 at 16 in. vertical and No. 5 at 15 in. horizontal

10.3.A. Reasonable choice: 12-in.-thick wall, No. 6 at 10 in. vertical at ¾ in. from inside face, No. 4 at 18 in. vertical at 2 in. from outside face, No. 4 at 13 horizontal in two layers (with both sets of verticals)

10.4.A. Possible choice: $w = 3$ ft, $h = 11$ in., $t = 9$ in., $A = 10$ in.; bars—(1) No. 3 at 18 in., (2) 3 No. 4, (3) No. 3 at 18 in., (4) 4 No. 4

INDEX